Prime Numbers,
The Holy Grail of Mathematics

This page was intentially left blank.

Prime numbers,
The Holy Grail of Mathematics

A Brief Introduction To Prime Numbers

Thamer Naouech

2 53 47 43 41 37 31 29 23
3   59 89 83 79 73   19
5   61 97 71   17
7   67   13
            11

Prime Numbers, The Holy Grail Of Mathematics

Copyright © 2020 Thamer Naouech

All rights reserved. No part of this publication may be reproduced, distributed, or transmitted in any form or by any means, including photocopying, recording, or other electronic or mechanical methods, without the prior written permission of the publisher, except in the case of brief quotations embodied in critical reviews and certain other noncommercial uses permitted by copyright law. For permission requests, email the author at the following email address: thamer.naouech@gmail.com .

ISBN : 9798839533677

First Edition, 2020.

This book is dedicated to my mom, dad, brother, and to my dear friends, the primes.

> I hope that...I have communicated a certain impression of the immense beauty of the prime numbers and the endless surprises which they have in store for us.

D. Zagier, *The first 50 million prime numbers*, The Mathematical Intelligencer (1977)

Table of contents

List of figures x
A note from the author xiv
Acknowledgments xvii
Introduction 1

1. Chapter 1 : Primes in nature 5
2. Chapter 2 : Primes and Aliens 11
3. Chapter 3 : Primes and cryptography 19
4. Chapter 4 : How many primes are there ? 30
5. Chapter 5 : The fundamental Theorem of Arithmetic 33
6. Chapter 6 : Sieving for Primes 38
   - Sieve of Eratosthenes 38
   - Sieve of Euler 42
7. Chapter 7 : The prime-counting function π 49
8. Chapter 8 : Euler's Totient function 58
   - Some properties of $\varphi(n)$ 59
   - Perfect totient numbers 63
   - Carmichael's totient function conjecture 64
   - Lehmer's Totient Problem 66
9. Chapter 9 : Mersenne Primes 67
   - The connection between Sophie Germain primes and Mersenne primes 69
   - Hunting for Mersenne primes 73
     - How to find Mersenne primes 76
     - Interesting facts about Mersenne primes 77

- Perfect numbers 78
10. **Chapter 10 : Pierre de Fermat** 85
    - Fermat's theorem on the sum of two squares 94
    - Fermat's Last Theorem 95
        - Θ Generalization of Fermat's Last Theorem 99
            - o Fermat-Catalan Conjecture 99
            - o Beal's Conjecture 100
    - Fermat's Little Theorem 100
    - Fermat Quotient 104
        - Θ Wieferich primes 105
            - o Wieferich numbers 107
    - Fermat Numbers 107
        - Θ Applications of Fermat Numbers 114
        - Θ Generalized Fermat Numbers 114
11. **Chapter 11 : The Riemann Hypothesis** 116
    - Riemann's work in number theory 121
12. **Chapter 12 : Primes in Arithmetic Progressions** 135
13. **Chapter 13 : Dirichlet's Theorem** 142
    - Dickson's conjecture: A Generalization of Dirichlet's Theorem 144
        - Θ Hypothesis H: A Generalization of Dickson's conjecture 145
14. **Chapter 14 : Formulas for Primes** 148
    - Mill s' Theorem 150
    - Euler's quadratic 152
15. **Chapter 15 : The Goldbach Conjecture** 154
    - Goldbach Partition and Goldbach Numbers 160
    - Related problems 163
    - Goldbach Conjecture in literature 164
16. **Chapter 16 : Bertrand's Postulate** 166
    - Related problems 172
        - Θ Legendre's conjecture 172
        - Θ Brocard's conjecture 174

Θ Andrica's conjecture	174
Θ Oppermann's conjecture	176
17. Chapter 17 : Wilson's Theorem	**178**
• Generalized Wilson's Theorem	181
• Wilson Primes	182
Θ Generalized Wilson Primes	182
18. Chapter 18 : The Twin Prime Conjecture	**184**
• Generalization	190
Θ De Polignac's Conjecture	190
19. Chapter 19 : The Ulam Spiral	**192**
• Variants: Klauber Triangle	199
References	**203**
Credits and references of figures and images	**250**
Appendix I: List of the first 1000 prime numbers	**254**
Index	**261**

List of figures

- Chapter 1 : Primes in nature
  - **Figure 1**: 46 chromosomes in a human cell, arranged in 23 pairs 5
  - **Figure 2**: A Periodical Cicada 6
  - **Figure 3**: Different Broods of Cicadas in the US 7
  - **Figure 4**: Graphical representation of the lifespan of different broods of cicada 10

- Chapter 2 : Primes and Aliens
  - **Figure 1**: The Golden Record cover shown with its extraterrestrial instructions 12
  - **Figure 2**: Arecibo radio telescope, Puerto Rico. 14
  - **Figure 3**: A representation of the Arecibo message 15
  - **Figure 4**: Explanation of the Arecibo message. 17

- Chapter 3 : Primes and cryptography
  - **Figure 1**: Caesar, Julius (100-44 Bc) 19
  - **Figure 2**: Caesar cipher with a shift of 3 20
  - **Figure 3**: Martin Hellman, center, and Whitfield Diffie, right, and Ralph Merkle (1977) 22
  - **Figure 4**: The number line 23
  - **Figure 5**: Numbers from 0 to 11 illustrated in a loop 24

- Chapter 5 : The fundamental Theorem of Arithmetic
  - **Figure 1**: Euclid of Alexandria 33
  - **Figure 2**: Carl Friedrich Gauss (1777 - 1855) 34

- **Chapter 6 : Sieving for Primes**
  - **Figure 1:** Eratosthenes of Cyrene (276-194 BC) 39
  - **Figure 2:** Sieve of Eratosthenes: even numbers crossed out 39
  - **Figure 3:** Sieve of Eratosthenes: even numbers and multiples of 3 crossed out crossed out 40
  - **Figure 4:** Sieve of Eratosthenes: all composite numbers are crossed out and only prime numbers are left 40
  - **Figure 5:** Leonhard Euler (1707 - 1783) 42

- **Chapter 7 : The prime-counting function π**
  - **Figure 1:** Graphical representation of the prime-counting function π(x) 51
  - **Figure 2:** Graphical representation of π(x) and x/ln(x) 53
  - **Figure 3:** Graphical representation of π(x), x/ln(x), and Li(x) 54

- **Chapter 8 : Euler's Totient function**
  - **Figure 1:** The graph of φ(n) when n ≤ 1000 59

- **Chapter 9 : Mersenne Primes**
  - **Figure 1:** Marin Mersenne (1588-1648) 68
  - **Figure 2:** Sophie Germain (1776 - 1831) 70
  - **Figure 3:** Derrick Henry Lehmer (1905-1991) 76
  - **Figure 4:** Édouard Lucas (1842-1891) 76
  - **Figure 5:** Tower of Hanoi 78
  - **Figure 6:** Niccolò Fontana Tartaglia (1499/1500 - 1557) 82

- **Chapter 10 : Pierre de Fermat**
  - Pierre de Fermat 85
  - **Figure 1:** The first 4 triangular numbers 90
  - **Figure 2:** The first 4 square numbers 91
  - **Figure 3:** Pentagonal numbers 91
  - **Figure 4:** A tetrahedron with side length 4 contains 20 spheres. 93

- o **Figure 5**: The 6th octahedral number: 146 magnetic balls, packed in the form of an octahedron with side length 6. 93
- o **Figure 6**: Andrew Wiles 96

- **Chapter 11 : The Riemann Hypothesis**
  - o Bernhard Riemann 116
  - o **Figure 1**: The first page of Bernhard Riemann's article concerning the number of primes that are less than a given magnitude. 117
  - o **Figure 2**: The spiral of Theodorus 123
  - o **Figure 3**: A metric relation in a right triangle 124
  - o **Figure 4**: Different sets of numbers 124
  - o **Figure 5**: Graphical representation of some complex numbers 127

- **Chapter 12 : Primes in Arithmetic Progressions**
  - o **Figure 1**: Edward Waring (1736 -1798) 138
  - o **Figure 3**: Terence Tao 138
  - o **Figure 3**: Ben Joseph Green 138

- **Chapter 15 : The Goldbach Conjecture**
  - o **Figure 1**: Letter from Goldbach to Euler, June 7th, 1742. .156
  - o **Figure 2**: Scatterplot of the number of Goldbach partitions G(n) for the first 1000 integers 162

- **Chapter 16 : Bertrand's Postulate**
  - o **Figure 1**: Pafnuty Lvovic (1821 – 1894) 169
  - o **Figure 2**: Joseph Louis François Bertrand (1822 – 1900).... 169
  - o **Figure 3**: Srinivasa Ramanujan (1887 – 1920) 169
  - o **Figure 4**: Paul Erdős (1913 -1996) 170
  - o **Figure 5**: The graphical representation of $A_n = \sqrt{p_{n+1}} - \sqrt{p_n}$, where p_n is the nth prime number, for for the first 100 integral values of n. 175

- Chapter 17 : Wilson's Theorem
    - **Figure 1**: Sir John Wilson (1741 – 1973) 179
    - **Figure 2**: Hasan Ibn al-Haytham (c. 965 – c. 1040) 180

- Chapter 19 : The Ulam Spiral
    - **Figure 1**: Stanisław Ulam 193
    - **Figure 2**: Recreation of Stanisław Ulam's scribbles 194
    - **Figure 3**: Ulam spiral of size 200×200 195
    - **Figure 4**: 150×150 Ulam spiral showing both prime and composite numbers. 198
    - **Figure 5**: Klauber's Triangle with 10 rows 200

A note from the author

Growing up, I was captivated by the charm of reading books and learning interesting facts about the world around me. However, there was a particular book that I would never forget. *"The Handbook of Mathematics"*. was a fascinating 600-page book talking about the nature of mathematics and how this scientific field is strongly connected to our lives. In the beginning, it was exciting because I finally knew the answers to a lot of questions that I asked my math teacher but he didn't answer, advising me to look for the answer on my own. So, I did. Then, at the age of 10, I was just learning about different kinds of numbers such as natural numbers, decimal numbers, rationals, ... and how to handle simple operations such as addition, substruction, multiplication, and division involving those numbers. I learned from that book how to "think deeply about math", even though the math I was doing was only about elementary counting. That book was a wonderful introduction to number theory whose purity I find enchanting. There is no doubt such introduction to number theory discussed prime numbers. That was the first time I learned about this interesting topic. Their definition was so easy and I understood it immediately. One of the childish ways I thought about primes is that they are divisible only by 1 and themselves because they don't trust any other number except themselves of course and 1 since it is the building-block of all positive integers. Silly, isn't it? However, I felt that there was something special about those numbers. They were different from others, but I did not know why. I had a deep feeling of fascination mixed with a passion that drove me to learn more about them. This passion kept increasing and finally got me to write this book.

For the previous 7 years, I have been studying prime numbers trying to understand what is special about them. I spend almost my entire free time reading undergraduate or even graduate textbooks and attending online lectures by great mathematicians in this field. At first, I didn't understand anything and there was a lot to learn starting from the mathematical jargon up to the essence of the formulae, the theorems, and the conjectures ... However, that feeling of confusion and frustration when I didn't grasp something could only encourage me to keep learning. Honestly, in the beginning, it was hard to find a book that was only about prime numbers and that was not a textbook, however, this did not prevent me from pursuing my passion. Reflecting on those days, I got a "crazy" idea. I decided to write a book for teenagers who are, like me, in love with prime numbers, hoping it will be a brief introduction to the holy grail of mathematics

I tried in this book to be as informative as possible and to make the concepts as clear as possible without getting into details that may require an advanced mathematical background. Nevertheless, theorems that are included here were followed by their proofs whenever I estimated the proof to be easy. If you are not a fan of proofs, you can skip those sections. However, I do not encourage you to do so. If you don't like proofs, please know that although they seem hard and complicated sometimes, they are the evidence that a certain statement is correct. Mathematics cannot evolve unless new theorems and methods are developed. We cannot claim that a theorem is true and use it unless we can prove mathematically that it is so.

"Why is it true and how can we prove it?" is at least as important a question as "What is true and how can we use it?", said **Gove Effingerin** in his *An Elementary Transition to Abstract Mathematics*.

Becoming familiar with proving theorems has a great impact on one's mathematical and logical skills. So, don't skip the proofs!

Who am I ?

My name is Thamer Naouech as you probably knew. When I wrote this book, I was a senior high school student majoring in Mathematics at the Elite High School of Kairouan, which is one of the most prestigious schools in Tunisia. I may not yet have a Ph.D. in number theory nor even an undergraduate degree in mathematics and I may not be the best to talk about prime numbers, however, it is only a strong passion that led me to write this book and to overcome all kinds of obstacles that I faced, for no reason but to share my genuine appreciation of prime numbers, that I consider as "the holy grail of mathematics". I tried to help middle or high school students in particular, and anyone in general through this brief introduction to one of the most fascinating topics. Writing this book took me three years that I spent searching for the best ways of explaining the content of this book and picking the most reliable resources that you will find in the References section.

Please do not hesitate to contact me directly on my email thamer.naouech@gmail.com for feedback.

Thamer Naouech

Acknowledgments

I thank my family that always supports me whenever I need help.
I am also truly indebted to Mr. **Kamel Troudi**, my dear math teacher who helped me through this journey. He has an exceptional ability to explain hard concepts in a way that not only makes it clear as crystal but also that makes you fond of it and eager to learn more.
Thanks also to all the people who believed in me and who kept encouraging me to do my best and to complete this project.

Introduction

At school, we all learned addition, subtraction, and multiplication. Yet, they may not be so exciting. However, when it comes to division, there is always something to discuss.

The first kind of division that a school pupil learns is Euclidian Division. According to the Euclidean division, dividing a by b means determining the unique two positive integers q and r verifying $a = b \times q + r$ such that $q \neq 0$ and $r < q$. q is called the quotient and r is the remainder. A deeper way of thinking of the Euclidian Division is by imagining the quotient as the number of bs that exist in a, and r as what is left when removing all the possible bs. Sometimes, when dividing a number a by another number b, the remainder is 0. In this case, we say that a is divisible by b or, equivalently, a is a multiple of b. In other words, there exists an integer k such that a = k×b.

Obviously, if b is strictly bigger than a, then a can never be divisible by b.

For example, if we divide 20 by 4, we get 5 as the quotient and 0 as a remainder. Thus, 20 is divisible by 4. It follows that 20 is also divisible by 5 because 4×5 = 20. Moreover, if we divide 20 by 10, we get 2 as the quotient and 0 as a remainder. However, if we divide it by 3, the quotient is 6 and the remainder is 2. Therefore, some numbers such as 2, 4, 5, and 10 divide 20, however, others such as 3 doesn't.

2 Introduction

Let's take another number, 19 for example. If we divide 19 by 2, we get 9 as the quotient and 1 as the remainder, by 3, the remainder is also 1, by 4, it is 3, ... If we try to divide 19 by all the positive integers less than it, we will always get a remainder. Therefore, except for 1 and 19, this number is not divisible by any other positive integer. Numbers with this property are called Prime Numbers.

Simply, a prime is a number strictly bigger than 1 that is only divisible by one and itself. This is simple and I am almost sure you have come across this at school. We all learned that for some special integers p, you can find no other integers that divide p but 1 and p itself.

In an increasing order, prime numbers are
2, 3, 5, 7, 11, 13, 17, 19, 23, 29, 31, 37, 41, 43, 47, 53, 59, 61, 67, 71, 73, 79, 83, 89, 97, 101, 103, 107, 109, 113, 127, 131, 137, 139, 149, 151, 157, 163, 167, 173, 179, 181, 191, 193, 197, 199, 211, 223, 227, 229, 233, 239, 241, 251, 257, 263, 269, ...

As you may have noticed, except 2, all prime numbers are odd which is intuitive because unless a number is 2, if it is divisible by 2, then by definition, it cannot be prime.

If a given number is divisible by another number different from 1 and itself, then it is called a composite number.

However, there exists a problem: to which category belong 0 and 1? You may think that this is easy since they are not primes, then they are composite numbers. This is not true, however. Indeed, the case of 0 and 1 is special. They are, conventionally, neither primes nor composites. Each one of them is its own category. So, to sum up, we have 0, 1, prime numbers, and composite numbers. It is worth mentioning that 1 used to be considered as a prime number, however, this is a convention that is no longer followed.

Historically, retrieved manuscripts such as the *Rhind Mathematical Papyrus* that dates to around 1550 BC, have shown that the Egyptians had known prime numbers. However, those special numbers were explicitly studied by Greek mathematicians especially **Euclid of Alexandria** around 300 BC.

Introduction

Prime numbers are an important topic in mathematics. As a matter of fact, they occupy a vast area of number theory which is a branch of mathematics studying the natural numbers. However, the charm of prime numbers goes beyond their simple definition. They appear among the natural numbers "*like invisible shining stars scattered throughout the endless numerical universe*"*, hiding a lot of secrets that mathematicians are trying their best to reveal.

In this book, I tried to cover some of the secrets that were revealed, and that are being revealed.

In the first chapter of this book, I talked about prime numbers from a different perspective. While prime numbers are a purely abstract mathematical object, it turned out that they exist in nature. They show up fascinatingly in the life cycle of some tiny insects called the Cicadas that used primes to maintain a "balanced life".

In the following two chapters, we will discuss some interesting applications of prime numbers. The first is indeed a funny one. Can you believe that primes were used to communicate with Aliens? The second application is more serious. You will understand how our "digital lives" depend on prime numbers. We will talk about modern cryptography and how prime numbers are used in the design of cryptographic systems.

Later, we will get deeper into the math behind primes learning how Euclid proved, around 2000 years ago, that there exist infinitely many prime numbers and how Eratosthenes discovered an interesting method to determine all prime numbers up to a given number, called the Sieve of Eratosthenes.

From the Greeks, we will get to recent work concerning primes starting from the 16th century and talking about Pierre de Fermat's outstanding achievements in mathematics in general and in the study of prime numbers in particular. Then, we will look at special kinds of prime numbers such as Mersenne primes.

When discussing famous conjectures involving primes, the Riemann Hypothesis stands out as a strong candidate. We will not only talk about the Riemann Hypothesis and the well-known Riemann Zeta function but also

4 Introduction

about many other conjectures and open problems such as the twin Prime conjecture, Goldbach conjecture, Dickson's conjecture, ... that are extremely important. Some conjectures are 100-year old, others were proposed 200 or 300 years ago, some were proved and some are still resisting.

After learning about those conjectures, you will certainly notice that many of them are easy to understand and don't require an advanced background in number theory, however, this apparent easiness hides an extremely difficult challenge for mathematicians. For instance, Twin prime conjecture and Goldbach Conjecture are in this last category. They are comprehendible effortlessly, but proving them has been a nightmare to the greatest number theorists.

It is important to mention that mathematics in general, and number theory in particular was revolutionized by the invention of computers. Indeed, checking numerically if a conjecture holds has become easier. What required years and maybe decades in the past is today done in a matter of minutes or even seconds. Thus, Computational number theory, an area of computational mathematics which is a new branch of math and computer science, arose and it has been developing methods and techniques to facilitate solving number theoretical problems using the computer.

Note:

> The words "number", "integer", and "natural number" are used interchangeably, unless otherwise indicated.

Chapter 1

Primes in nature

It is such a wonderful moment when you reflect on the flowers, the animals, the rivers,... trying to see through nature and to reveal its secrets that never stop fascinating us.

Mathematical patterns are frequent in nature and just listing them may require an entire book Yet, I will talk only about a few of them, even though it will be a pleasure for me to discuss all of them because they are truly fascinating.

23! It is an interesting number, at least from a mathematical perspective. It is special because it is prime.

Try to think, right now, what this number may represent.

Remember that biology course you may have taken at school when you learned that human beings have 23 pairs of chromosomes making their total 46. [1]

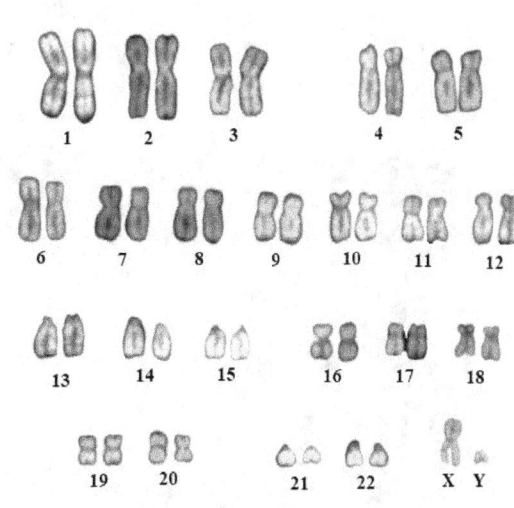

Figure 1: 46 chromosomes in a human cell, arranged in 23 pairs.

6   Primes in nature

Researchers have proven that our primitive ancestors had 24 pairs of chromosomes. However, around 66 million years ago there was a head-to-head diffusion between two chromosomes reducing their total number by 1. [2] May this fusion be what made us who we are?

To date, it is not known what is special about having a prime number of pairs of chromosomes.

However, this may not be the kind of pattern we are looking for.

Let's continue our search for prime numbers in nature.

Do you like insects? Maybe yes, maybe no. However, a certain kind (technically speaking, a certain family) of them is worth learning about.

Cicadas are little insects with prominent eyes and short antennae. They are medium to large in size, ranging from 0.8 to 2 inches (2 to 5 cm), and have an exceptional song that you may recognize easily. Some species of cicadas spend most of their or that lives underground and emerge only after a decade, a behavior described by scientists as an adaptation strategy.   [3] Cicadas have the largest lifespan among all insects, but only one genus(a biological classification meaning a species, a kind, …) among the 3000 known is more interesting than others, at least for us, even though they may look scary and weird coming out of the ground with the red beachy eyes. This genus is called the Periodical cicadas or Magicicada.

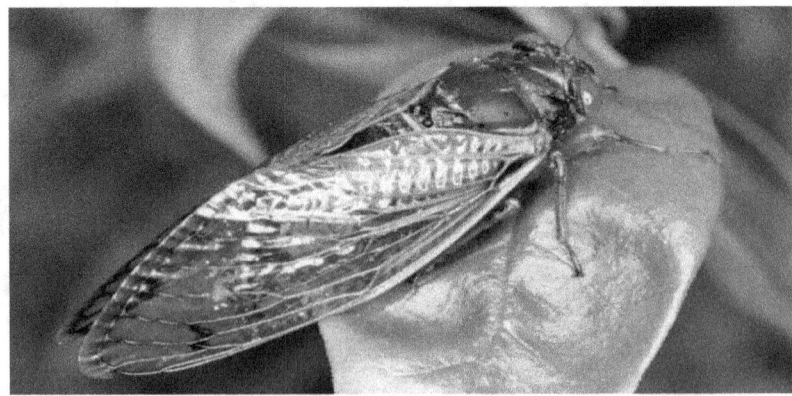

Figure 2: A Periodical Cicada

The US is home to 15 species of Magicicadas. Among those, 12 have a 17-year- lifespan, and the other three can live for 13 years. [4]

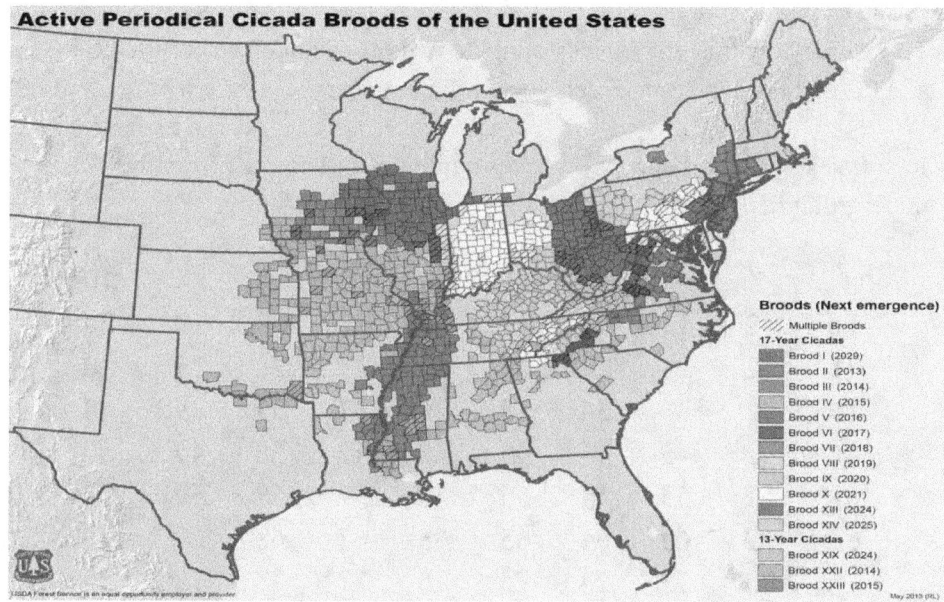

Figure 3: Different Broods of Cicadas in the US

To date, we do not understand thoroughly how this cycle works. However, we are sure that they emerge from the underground in the spring of their 13th or 17th year, depending on the brood, to achieve their final form. Then they start singing their famous and last song. But what is so interesting about them is their lifespans. As you may have noticed, they are prime numbers and you are about to know why this is a vital necessity. [4+5]

Experts in this field have shown that if two or more species of cicadas appear at the same time and they interbreed, the whole synchronized cycle of each group will break, and thus causing many problems. Therefore, the end of the lifespan of two broods shouldn't overlap, at least not frequently, and it is also important that they do not appear at the same place at the same time, to assure the integrity of their genes, and more importantly, their entire existence.

8 Primes in nature

The way those broods of cicadas maintain this balance is genuinely amazing.

In fact, this is based on a simple arithmetic concept.

To explain how this works, let's imagine two broods of cicadas called C1 and C2 that appear, respectively, every 4 and 6years.

In 100 years, C1 will appear 25 times while C2 will emerge from the underground 16 times.

1	2	3	4	5	6	7	8	9	10
11	12	13	14	15	16	17	18	19	20
21	22	23	24	25	26	27	28	29	30
31	32	33	34	35	36	37	38	39	40
41	42	43	44	45	46	47	48	49	50
51	52	53	54	55	56	57	58	59	60
61	62	63	64	65	66	67	68	69	70
71	72	73	74	75	76	77	78	79	80
81	82	83	84	85	86	87	88	89	90
91	92	93	94	95	96	97	98	99	100

The yellow and blue squares represent the years of emergence of C1 and C2, respectively.
The green squares show the overlapping years.

In some specific years, the two broods of cicadas appear together and this happens, in our case, once every 12 years or almost 8 times per century. However, this is dangerous because it increases the probability that the two broods will interbreed.

Those specific years, such as 12, 24, 36, …, in which the two broods of cicadas appear together are, as you may have noticed, multiples of both 4 and 6 and thus they are multiples of the **least common multiple** of 4 and 6 which is 12.

> **Notes:**
>
> - The **least common multiple**, also known as the **lowest common multiple**, of two non-zero integers a and b is the smallest positive integer that is divisible by both *a* and *b*. It is usually denoted by **lcm(*a*, *b*)**.
> - The **greatest common divisor** of two non-zero integers a and b is the greatest positive integer that divides both of them. It is usually denoted by **gcd(*a*, *b*)**.
> - The gcd and the lcm of two non-zero integers a and b are connected by the following expression:
> $$gcd(a,b) \times lcm(a,b) = a \times b$$

So, if you think about this for a while, you will figure out that the best way to lessen the number of times both broods of cicada appear together is by taking two numbers with an lcm as large as possible. And there comes nature's trick!

Based on the identity stated above, the lcm is maximum when the gcd is minimum. Or the gcd of two integers a and b is always bigger or equal to 1. So, the minimum value the gcd can take is 1. Therefore, in this case, the lcm will be equal to the product of a and b.

For the gcd to be equal to 1, a and b have to be relatively prime.

> Two integers are said to be **coprime** or **relatively prime** if and only if their gcd equals one.

So, let's check if this works for 13 and 17. According to our calculations, the two broods of cicada will appear together once every 221 years or 4 to 5 times per millennium.

You may be wondering why 17 and 13 specifically? Why do we care about them being prime or not? What we need here are just too numbers that are relatively prime such as 12 and 13 for example or 7 and 9, ...

This is a very interesting question!

Primes in nature

It turns out that cicadas' predator populations also move in cycles. Depending on the abundance of the food and the competition with other predators, their lifespan can shift a little bit to fit the patterns of their preys. If the cicadas chose a 12-year cycle, then every predator of 2,3,4 or 6-year cycle can also adjust their lifespan.

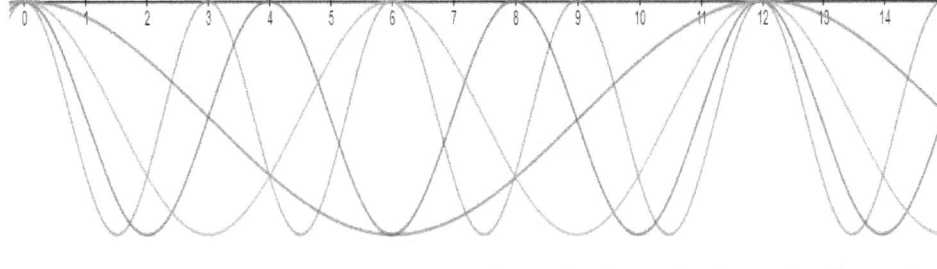

3-year cycle   4-year cycle   6-year cycle   12-year-cycle

Figure 4: Graphical representation of the lifespan of different broods of cicada.

However, by emerging every 13 or 17 years, cicadas minimize those deadly coincidences. A predator with a 5-year cycle, for instance, will only align with cicadas every 85 years.
 Therefore, choosing the lifespan to be a prime number is an optimal solution for cicadas because it's not divisible by any smaller number (except one, of course). [6]
But how can this be possible? Can those tiny insects understand prime numbers and use them and designing their lifespans?
Actually, no. As far as we can tell, cicadas cannot understand math, let alone prime numbers. However, this strategy had been shaped over at least 4 million years. [7]

Chapter 2

Primes and Aliens

Do aliens exist? Probably. Who knows. But let's think about it deeply. Considering the vastness of the universe, maybe somewhere beyond the nebulas, in other galaxies some intelligent creatures might exist.

In fact, for decades, we have been patiently scanning the sky for any sign of intelligent life.

However, there might be a small problem. Assuming that an intelligent life beyond earth exists and that we try to communicate with them, what language will we use?

Will we use some of the earth's languages? This does not seem like a reasonable decision, even though earth languages were used before in a message that was sent to interstellar space. Greetings from Earth-people in 55 languages were sent on a phonograph record: a 12-inch gold-plated copper disk that was carried by the Voyager spacecraft (1977). [1]

12 Primes and Aliens

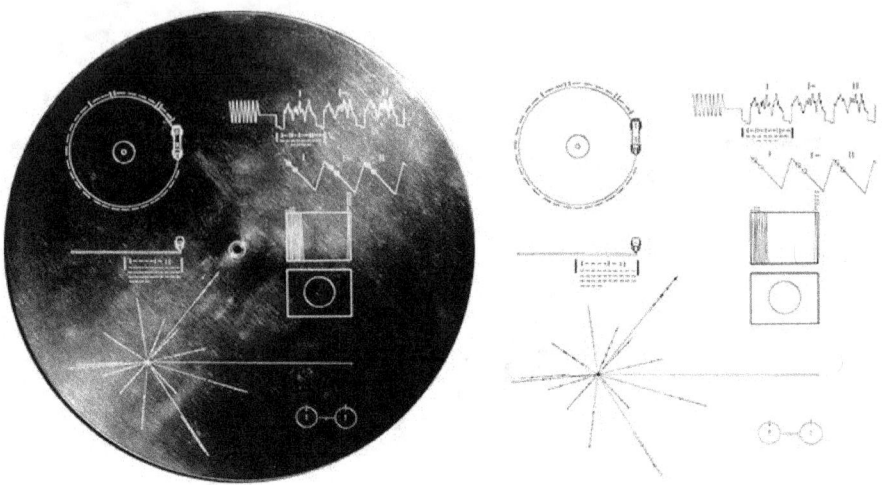

Figure 1: The Golden Record cover shown with its extraterrestrial instructions.

The chances that extraterrestrial beings -if they exist- speak English or French or German or any other language on earth are very small.

In his book, published in 1960 and titled *"Lincos: Design of a language for cosmic intercourse, Part 1"*, the German mathematician and philosopher **Hans Freudenthal** suggested a new language, that he called "Lincos", to be used in the communication with extraterrestrial intelligence.

[2]

Each symbol in Lincos is defined by symbols that came before it so that you don't have to know anything apart from pure logic to understand it. It is supposed that we will first transmit a Lincos dictionary to teach Aliens this language. As Dr. Freudenthal included in his book, we will teach them many concepts ranging from simple arithmetic to time to describing human behaviors ...

[3]

In 2000, Dr. **Yvan Dutil** and **Stephane Dumas** from the Defence Research Establishment Valcartier in Canada used Lincos to encode a message that was sent to outer space from Evpatoria transmitter in Ukraine.

[4]

In the 1970s, in another attempt to solve this problem, American astronomer **Frank Drake** designed an astounding method for extraterrestrial communications involving prime numbers.

[5+6]

Primes and Aliens

A message encoded using this technique is just a sequence of dots and dashes like in the following example.

```
_ _ _ _ _ _ _ . _ _ _ _ _ _ _ _ _ _ _ . . . _ _ . _ _ _ _ . . . . .
_ . . _ _ . . . . . . . . . . . _ _ _ _ . _ _ _ . . . _ _ _ . _
_ _ . . . . . _ _ . _ _ _
```

Well, let's assume, for a moment, that you are an alien who has just received a series of dots and dashes that seems like a message. For example, let's say you received a sequence of 77 dots and dash. Out of curiosity or just because you are doing your job, you tried to understand its meaning. So, you started to play around with it.

After noticing that 77 can be factored as 7×11 or 11×7, you got a brilliant idea. You rearranged the dots and dashes into a rectangular form. So, you drew a table with 11 columns and 7 rows and another with 11 rows and 7 columns.

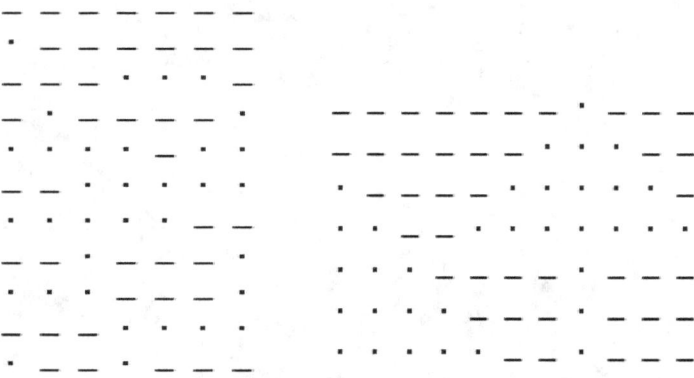

Clever by nature, you noticed the first array (on the left) is nonsense because, apparently, dots and dashes are randomly distributed. However, in the second, there might be a pattern. In the left part of it, there is a right triangle. Maybe whoever sent this message knows the " Pythagoralian Theorem ". Or maybe he/she/it is just listing numbers from one to five denoted by dots and starting from the 3rd row. And what is that shape next to the triangle? Is that an arrow? Or is it a Spruce tree?

As you may have noticed, dots and dashes can be interpreted as pixels, those tiny squares that make the screen of the monitor or the smartphone, ... And

14 Primes and Aliens

thus, we can use this method to send images and maybe entire videos. We may not be able to transmit high-resolution images, but it will be enough to send simple information telling the story of who we are.

As we have discussed, this method converts a simple message to a sequence of dots and dashes that are supposed to be arranged into a rectangular form. And there comes the trick. The total length of the sequence has to be the product of two prime numbers to reduce the number of possible arrays to two. Nevertheless, we can do better by choosing the length to be the square of a prime.

This technique was used indeed to encode a message broadcasted in November 1974 by the Arecibo radio telescope, installed in Puerto Rico, to globular star cluster M13. This message has been called The Arecibo Message. [7]

Figure 2: Arecibo radio telescope, Puerto Rico.

This message contained 1679 bits*, that is, 23 × 73. When arranged in a rectangular form, we get the following picture:

> * Bits are 1s and 0s. A "0" is represented by an "off" radio pulse, while a "1" is represented by an "on" radio pulse.

Primes and Aliens 15

Figure 3: A representation of the Arecibo message

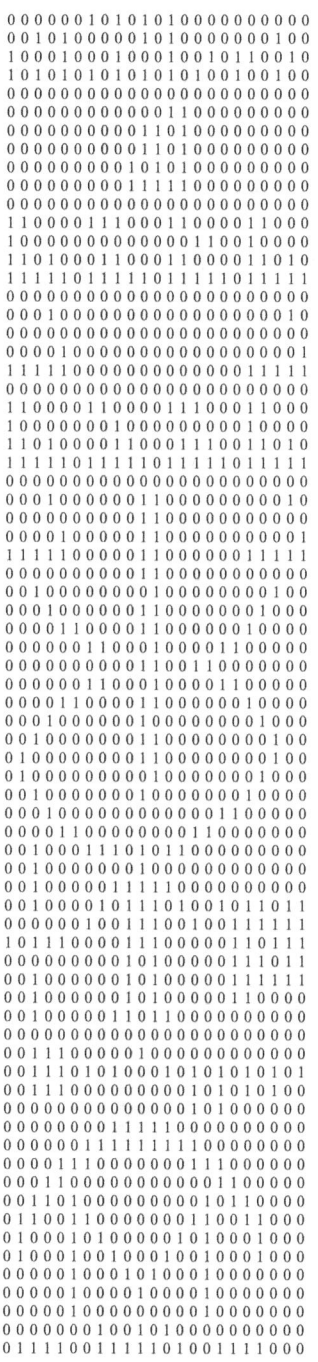

The original Arecibo message represented in ones and zeros.

[8]

16   Primes and Aliens

The Arecibo message contained basic information such as the numbers from 1 to 10, the atomic number of Hydrogen, Carbon, Nitrogen, Oxygen, and Phosphorus which are the most important elements for life on Earth, bases of nucleotides in our DNA, a picture of a human being, representation of the solar system, details about the Arecibo Observatory, etc.
The figure on the next page explains, in more details, the components of this interesting message. [7]

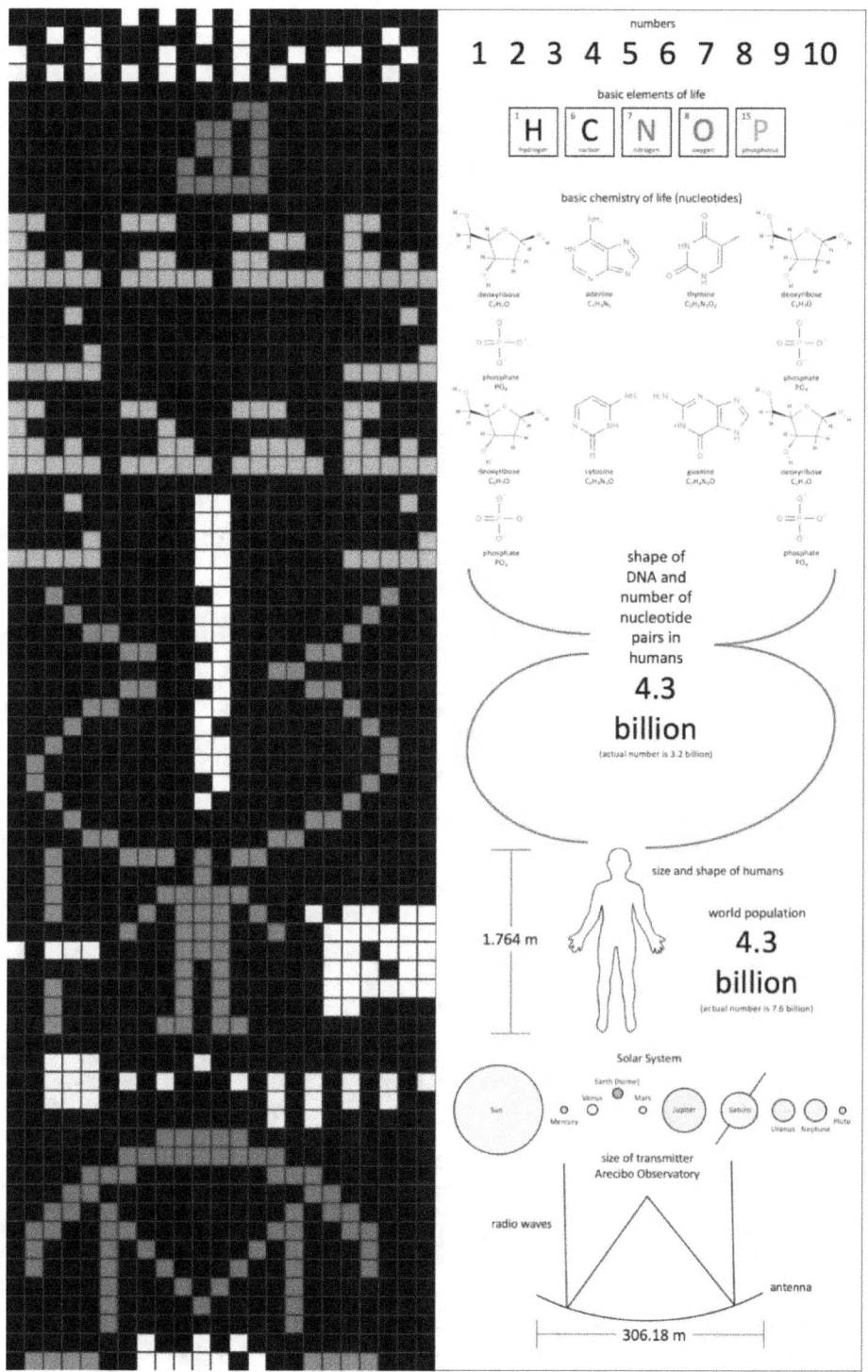

Figure 4: Explanation of the Arecibo message.

18 Primes and Aliens

It is such a brilliant idea to use prime numbers in designing a communication system with extraterrestrials because if there exists an intelligent civilization somewhere in the universe and it is developed enough to build a radio telescope and to receive our message, it should know prime numbers and maybe better than we do.

Chapter - 3

Primes and cryptography

For hundreds of years kings and generals, and recently governmental agencies and presidents have been using different ways of communication to send or to receive classified information. Therefore code makers have been designing new techniques to hide messages, a process called **encryption.** Simply, encrypting a message means finding a way to disguise that message so that only the intended recipient can "read" it.

While some nations have been trying to find new and efficient ways of encryption, rival nations have been trying to intercept and break their enemy's encrypted messages. For that reason, professional code breakers were hired.
The first known encryption method is called Caesar's Cipher, after the Roman Emperor **Julius Caesar.**

To encrypt a message(supposed in English) using this method, you have to define a key which is in our case a number between 1 and 26. Let's say you chose 7. Now, you take

Figure 2:
Caesar, Julius
(100-44 Bc)

your original message and substitute each of its letters with the letter that is seven places further down the alphabet.

This technique is also called Caesar Shift Cipher and the reason for this is obvious since you are shifting the plane text's alphabet (the original message) by, in this example, 7 positions to get the Cipher alphabet.

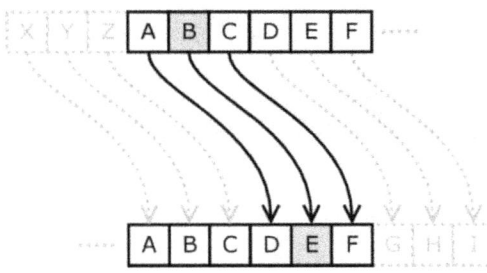

Figure 2: Caesar cipher with a shift of 3 (A becoming D when encrypting, and D becoming A when decrypting)

However, this method is inefficient because it is easy to try all the 26 possibilities even by hand and you will ultimately reach the original message.

Therefore, throughout history, code makers have designed new and more efficient methods of encryption like Viginère, Gronsfeld, and Baconian Cipher ….

However, all those methods are obsolete when compared to the ENIGMA, an encryption machine that was used by the Germans in World War II and it was a nightmare to the Allies [1]. Fortunately, it was finally broken thanks to many techniques developed by English mathematician and cryptanalyst **Alan Turing**.

Turing's work was the first step toward designing the programmable calculator, the ancestor of our modern computer. [2]

During the 1960s, computers became cheaper and more powerful. Therefore, businesses could afford them to be used in various tasks. For instance, banks started to use them in money transactions, governments used them for communication and military-related tasks …etc. In short, computers were used to send and to receive information. However, transmitting a piece of

Primes and cryptography

information such as bank credentials is very dangerous because it may be easily intercepted by eavesdroppers.

So, banks had to use encryption. Many banks used encryption software to encrypt the credential data of their clients who had to have a copy of that software on their computers to decrypt the messages received from the bank. But there was a problem.

How can the bank and the customer have the same key to encrypt or decrypt the messages they exchange? In other words, how the customers are supposed to know the key their bank used in the encryption?

Going to the bank each time the customer wants to make a transaction doesn't seem like a practical solution. Using the phone is also insecure because what if someone is taping the wire?

As an attempt to solve this riddle, some banks hired trustworthy couriers to distribute, in person, the keys to the clients around the world. But this wasn't a reasonable solution.

Indeed, this problem called the Key Distribution problem wasn't new. It had been challenging cryptographers throughout history.

In the 1960s, the US Department Of Defense started a new organization called the Advanced Research Project Agency (ARPA) that would develop a network of inter-connected computers that was a primitive version of what we know today as the Internet.

However, the Key Distribution problem represented a huge weakness in this new project.

Many thought this problem was unsolvable until the mid-1970s when a brilliant team of mathematicians came out with an astonishing solution.

In 1976, **Whitefield Diffie**, an enthusiastic cryptographer, **Martin Hellmann**, a professor from Stanford University, and **Ralph Merkle**, a cryptographer and a researcher, published a scientific paper explaining their simple, yet mind-blowing discovery. [3]

22 Primes and cryptography

This solution in simple words goes like this:

Alice wants to send Bob a message. So she puts it in a box and closes it using a padlock and a key. Then, she sends it to Bob via the mail. Remember, only Alice has the key that can open the box.
So, Bob won't be able to open it. But here comes the brilliant trick! Bob won't open the box, instead, he will close it again using his padlock and then resend it to Alice. When she receives it, she will remove her padlock leaving only Bob's, and send it back to him. Thus, Bob will be able to remove his padlock using his key and read Alice's message. Brilliant, isn't it?

Figure 3: Martin Hellman, center, and Whitfield Diffie, right, and Ralph Merkle (1977)

     This solution is a breakthrough in cryptography since it created a new kind of ciphers: the Asymmetric Cipher. This kind of ciphers uses 2 keys: one encryption and the other for decryption.
It is important to mention that Caesar, Viginère, and Baconian Cipher are called symmetric ciphers since the same key that is used to encrypt the message is used to decrypt it.

Even though this method was a revolutionary solution to the key distribution problem, it was only theoretical with no real application.
Nevertheless, it helped cryptographers think about this problem differently and thus a new updated version Diffie-Hellmann's solution came out.
In fact, instead of sending and resending and re-resending the box containing the message, Alice can use a brilliant trick. She can make many copies of her padlock and send them to all the mail offices around the world. So whenever Bob wants to send Alice a message, he can go to the nearest mail office, asks

for Alice's padlock, and use it to close the box that contains his message. Finally, he sends it to her.

Technically, Alice has created a **public key** (the padlock) that will be used by anyone who wants to send her a message, and a **private key** which is the key she uses to decrypt the messages she will receive (her padlock's key). However, such a method cannot be implemented in real life unless we can find a mathematical function that has this property.

> In mathematics and computer science, there are two kinds of functions: one-way and two-way functions. The two-way function is easy to do and easy to undo.
>
> For example, think of $f(x) = x^2$. To undo what this function did, we can use its reciprocal function which is: $f^{-1}(x) = \sqrt{x}$

The team of Diffie and Hellmann was trying to find that mysterious function. They were not interested in the two-way function because it is easy to break, but instead, they focused their attention on the one-way functions because it is almost impossible to reverse them.

A one-way function is a function whose output is easy to compute but it is very hard to determine the input given the output unless you have a specific piece of information that you may think of as the key of the padlock in the scenario we discussed earlier.

One-way functions often use modular arithmetic which is a different kind of arithmetic.

If you are familiar with modular arithmetic, you can skip the following brief introduction.

> In the arithmetic we are familiar with, numbers can be arranged in a linear form using the number line.
>
>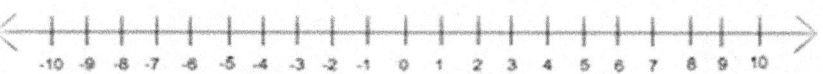
>
> Figure 4: The number line

24   Primes and cryptography

However, in modular arithmetic, you can imagine the numbers as arranged in a circular form or a loop like the numbers in the clock. They wrap around when reaching a given number.

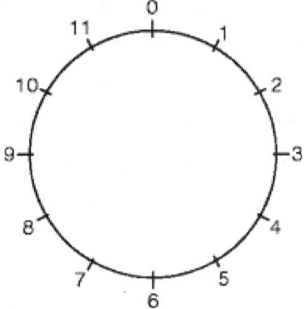

Figure 5:

Numbers from 0 to 11 illustrated in a loop (Similar to the numbers on a mechanical 12-hour wall clock)

Believe it or not, we are all familiar with modular arithmetic because it is used in our daily life, especially for those who use frequently mechanical watches.

First of all, let's agree about using the 24-hour time format.

Let's imagine the following situation. You are watching your mechanical watch which indicates 12 hours and 45 minutes. 15 minutes later, the time will be 13:00.

However, it is 1:00 according to your watch. Is the time indicated by your watch wrong? No, it isn't. Your watch is telling you that, right now, the time is 1 hour after 12 o'clock. It is 1 pm.

What was done is no more than a simple Euclidean division. We divided the actual time which is in this case 13 o'clock by 12 and took the remainder. This is simply what modular arithmetic is about. (Of course, it is more complicated than that when you study it more deeply, but it remains one of the most important and most interesting mathematical tools)

When divided by 12, 13 and 1 have the same remainder 1. In modular arithmetic, 13 and 1 are said to be congruent modulo 12.

More generally, if two integers a and b have the same remainder upon division by an integer n, i.e., their difference is divisible by n, then they are said to be congruent modulo n. n is called the modulus. This is denoted by

$$a \equiv b \pmod{n}$$

In the example we discussed above, the modulus n is 12.

Remark:

You can think of the congruence sign "≡" as an equal sign "=". However, the former is used only in modular arithmetic to denote the special modulo operation.

For example, 43 and 7 are said to be congruent modulo 12 because both of them have the same remainder, 7, when divided by 12.
$43 \equiv 12 \times 3 + 7 \equiv 7 \pmod{12}$.

We have also $43 \equiv 20 \times 2 + 3 \equiv 3 \pmod{20}$.

Properties of addition, multiplication, and exponentiation in modular arithmetic:

- Reflexivity: $a \equiv a \pmod{n}$

- Symmetry: $a \equiv b \pmod{n}$ if $b \equiv a \pmod{n}$ for all a, b, and n.

- Transitivity: If $a \equiv b \pmod{n}$ and $b \equiv c \pmod{n}$, then $a \equiv c \pmod{n}$

- Compatibility with addition and subtraction:
  If $a \equiv b \pmod{n}$ and $c \equiv d \pmod{n}$, then $a \pm c \equiv b \pm d \pmod{n}$

- If $a + b = c$, then $a + b \equiv c \pmod{n}$

- If $a \equiv b \pmod{n}$ and c is an integer, then $a \pm c \equiv b \pm c \pmod{n}$

Primes and cryptography

> ***Remark***
> Note that this is just a superficial introduction to modular arithmetic. However, if you are not familiar with this notion, I do encourage you to learn more about it, online or from textbooks, since it is an important tool in mathematics.

Let's go back to our encryption problem. As you may remember, we were talking about how many mathematicians and cryptographers were trying to find that puzzling one-way function that is easy to do but is hard to undo without using an additional specific piece of information.

While the team of Diffie and Hellmann was trying so hard to find it, another team from MIT Laboratory Of Computer Science joined the race. This team was composed of **Ron Rivest**, **Adi Shamir**, two computer scientists, and **Leonard Adelmann**, a mathematician.

They spent almost a year coming out with new ideas and approaches to tackle this problem, but all of them were flawed.

Nevertheless, their efforts soon paid off. One night, in April 1977, Rivest, while skimming over a math textbook, had an idea which turned out to be the best he ever had. All of a sudden, the solution simply popped up into his mind. [4]

He spent that whole night writing the paper that is considered, nowadays, as a breakthrough in computer science and modern cryptography. That paper contained a detailed description of a new cryptosystem. The new algorithm they invented was called RSA. This acronym is the initial letters of the surnames of Ron Rivest, Adi Shamir, and Leonard Adleman.

Explaining how this algorithm works requires an advanced background in mathematics and computer science.

However, in simple words, Alice has to create a public key and a private key which are analogous to the padlock and its key in our previous example. Then, she will publish her public key on the internet so that everyone can access it, keeping her private key secret. So when Bob wants to send her a message, he will search for her public key and use it, in addition to the new one-way function, to encrypt his message.

Primes and cryptography

> You may be wondering how a mathematical function can deal with a text composed of characters. The answer is really simple: all the characters (alphabet, numbers, and symbols such as ''@'', ''#'' ...) are represented in the computer as a sequence of binary digits (0s and 1s). So in encryption, as well as in decryption, the computer uses the binary representation of each character in the plain text as an input of the one-way function whose numerical output is converted into another character. And thus, character by character, you get the encrypted text.

Technically, Alice's public key is composed of two numbers. Let's call them x and e. What is special about x is that it is the product of two large prime numbers. This condition is essential for the cryptosystem to work efficiently.

To create her public key, Alice has to choose and multiply two large prime numbers. The product is part of the public key while the two prime numbers are part of the private key that has to be kept secret.
Suppose that Alice picked two prime numbers p = 19 and q = 31.
Thus, x = 19 × 31 = 589. Then, she picks another number, e, such that e and (p-1) × (q-1) are coprime. For example, let e = 23
After getting Alice's public key, Bob has to convert his message into a numerical value. For the sake of simplicity, we will assume that his message is a single letter. Let it be "P " and we will use its rank in the alphabet, which is 16, as ist corresponding numerical value.

Bob will encrypt his message using Alice's public key and the following one-way function:

$$C \equiv O^e \ (\bmod\ x)$$

where C is the ciphertext, in this case, the rank of the encrypted letter, and O stands for the original text, in our example, the rank of the original letter which is P.
So, $C \equiv \mathbf{16^{23}} \equiv 4 \ (\bmod\ 589)$

Thus, the encrypted message that will be sent to Alice is the 4th letter in the alphabet, i.e., D.

Primes and cryptography

When Alice receives the message, she will be able to decipher it using her decryption key, d, which can be calculated using the two prime numbers she chose initially by applying the following formula:

$$e \times d \equiv 1 \ (\bmod \ (p-1)(q-1) \ )$$

Hence, in this example, $d \times e \equiv 23 \times d \equiv 1 \ (\bmod \ 540 \ )$. Therefore, $d = 47$.

Remark:
>d is called the inverse of e modulo (p-1)(q-1) and it can be calculated using a straightforward method. However, it will be up to the curious reader to learn more about it.

By plugging Bob's interpreted message which is 4 and her decryption key into the following formula, she will get the original message.

$$O \equiv C^d \ (\bmod \ x \ )$$

In our case, $O \equiv 4^{47} \equiv 16 \ (\bmod \ 589 \ )$.
And Voila! Alice can then deduce that Bob's original message is "H". [5]

When an eavesdropper intercepts the transmission of the message, he/she will get only the encrypted sequence of nonsense. Even though he/she knows the function and the public key (x and e) that were used in the encryption process, he/she will not be able to decrypt it because he/she doesn't know the prime numbers that were initially used to create the public key.
You may have thought about factorizing x, the number used as the public key to determine the initial 2 primes. This may be simple when the number to be factorized is 119 or even 12591. Using a calculator, this won't take more than some minutes. However, it is a matter of some milliseconds when using a computer. But the larger the initial primes, the harder the factorization of their product is.

Here is a challenge for you: try to factorize this number:
132 153 274 734 801 909 554 131

Primes and cryptography

Sometimes, the predicted time of the factorization of the product of 2 prime numbers may be bigger than the age of our universe!
This is because we don't have a quick way to factorize numbers. The RSA cryptosystem is based on this fact. So, whenever we use it, we try to choose the two prime numbers as big as possible to make sure that even if someone is crazy enough to try to factorize the product, it will take him hundreds if not thousands of years. By the time he gets to the answer, if he did, the message will be useless.
At present, the minimum number of digits of the public key used in data transmission is 617. (2048-bits long, thus the initial primes are 1024-bits long each).
However, in the transmission of highly classified information, the initial primes are 2048-each making the public key 4096-bits long. [6]

* * *

It is really a wonderful moment when we reflect on these special numbers on which our modern life sensitively depends.
However, what if one day we discovered a quick and powerful way of factorization? What can happen to our systems, our economies, and most importantly, our privacy? Will our world collapse? Maybe and maybe not, hopefully not. Don't you think it's high time we started thinking of a more powerful cryptosystem?

Chapter 4

How many primes are there ?

This is an interesting question. Let's start counting. However, had you done so, you won't get to an end. This is because there is an infinite number of primes. Greek mathematician **Euclid** is one of the first mathematicians who studied prime numbers and he asked this question. Indeed, he managed to prove that there exist infinitely many of them. This was included in Book IX of his legendary *Elements*. (Proposition n° 20)[1]

The proof of this proposition is easy. It is done by the method of proof by contradiction.

Proving a mathematical statement using this method consists of supposing it does not hold, i.e., this statement is false, and then showing that this hypothesis (the statement being false) cannot be true, i.e., so, you can conclude that the original statement is true.

By way of contradiction, let's supposed that the number of prime numbers is finite, that is, there exist only k prime numbers and let all the prime numbers be $p_1, p_2, p_3, p_4, \ldots, p_{k-1}, p_k$, such that $p_1 < p_2 < p_3 < p_4 < \ldots < p_{k-1} < p_k$, where p_k is the largest prime number.

Now, let's consider the number

$$Q = p_1 \times p_2 \times p_3 \times \ldots \times p_{k-1} \times p_k + 1.$$

This number is not divisible by any of p_i, $1 \leq i \leq k$.

Even though this last statement is intuitive, we can prove it mathematically. By way of contradiction let's assume that there exist a prime number $p_r \leq p_k$ that divides

$$Q = p_1 \times p_2 \times p_3 \times ... \times p_{k-1} \times p_k + 1.$$

Or, $p_1 \times p_2 \times p_3 \times ... \times p_r \times ... \times p_{k-1} \times p_k$, i.e., $Q - 1$, is a multiple of p_r, and thus, divisible by p_r.

Using a basic result in arithmetic*, and given that p_r divides both Q and Q - 1, then we can deduce that p_r must divide the difference of those two numbers. Thus, it divides $Q - (Q - 1) = 1$. But this is a contradiction because p is a prime number and no prime number divides 1.

> *One of the fundamental results in arithmetic states that if a, b, and c are three integers such that c divides both a and b, then c must divide any linear combination of a and b, i.e., c divides $\alpha \times a + \beta \times b$, where α and β are integers (α and β they can be positive, negative, or even 0). In the proof above, a special case of the latter expression is used, where a = Q, b = Q – 1, $\alpha = 1$, and $\beta = -1$.

Hence, either Q is prime or it is divisible by a prime bigger than p_k. The first case means that this number is prime and it is greater than p_k. In both cases, a prime number bigger than p_k exist. But this is absurd because we assumed that p_k is the largest prime number. Thus, we can conclude that our initial hypothesis is false, which is to say that there exist infinitely many prime numbers.

■ [2]

In 2006, Dr. **Filip Saidac**, an associate professor of mathematics at the University Of North Carolina, USA, published a new proof of the infinite number of prime numbers. [2]

32   How many primes are there ?

His easy and intuitive proof goes as follows.

Let n be an arbitrary integer greater than 1. Consider the numbers n and n+1. Each one of n and n+1 has at least one prime factor. Based on a demonstration we did previously, those two consecutive integers don't share any common prime factor. Thus, their product has at least two different prime factors. (For example, let's consider 3 and 4. They are consecutive, thus, coprime. Their product $3 \times 4 = 2^2 \times 3$ has two different prime factors which are 2 and 3).

Similarly, n(n+1) and n(n+1)+1 are two consecutive integers whose product has at least three different prime factors. Repeating this process infinitely will yield, each time, a new prime number. Thus, the number of prime numbers is infinite. [2]

Chapter 5

The Fundamental Theorem of Arithmetic

If a number be the least that is measured by prime numbers, it will not be measured by any other prime number except those originally measuring it.

— Euclid, Elements Book IX, Proposition 14

Any number either is prime or is measured by some prime number.

— Euclid, Elements Book VII, Proposition 32

[1]

Figure 1:

Euclid of Alexandria

34 The fundamental Theorem of Arithmetic

The Fundamental Theorem of Arithmetic, as the name states, is one of the most essential theorems in number theory.

This theorem states that every integer n greater than or equal to 2, either is prime itself or can be represented as a unique product of prime numbers.

In other words, if n is an integer such that n ≥ 2, then

$$n = p_1^{\alpha_1} \times p_2^{\alpha_2} \times p_3^{\alpha_3} \times p_4^{\alpha_4} \times ... \times p_k^{\alpha_k},$$

where p_i are distinct prime numbers such that $p_1 \leq p_2 \leq p_3 \leq ... \leq p_k$, $1 \leq i \leq k$, and $\alpha_i \geq 1$. This representation is unique.

Note:

> Please note that rearranging the prime numbers doesn't count. What is important is the prime numbers themselves and their exponents. We count the factorizations 2×7×11, 7×2×11, and 11×7×2 as the same, for instance.

Writing n as the product of prime numbers is also called the prime decomposition of n.

It will make total sense to you to know that this theorem is also called the **unique factorization theorem** and sometimes the **unique-prime-factorization theorem**. This theorem may explain why 1 is not considered as a prime number because any product of prime numbers will be the same when multiplied by any number of 1s.

A different version of the fundamental theorem of arithmetic was stated for the first time by **Euclid Of Alexandria**. [1+2] However, the first modern statement of this theorem was included in **Carl Friedrich Gauss**' famous *Disquisitiones Arithmeticae*. Gauss did not only

Figure 2:
Carl Friedrich Gauss
(1777 - 1855)

The fundamental Theorem of Arithmetic 35

reformulate this theorem and prove it differently but also he generalized it. However, explaining Gauss' work is beyond the scope of this book. (It requires advanced knowledge in ring theory). [3]

The following part will be dedicated to a simple proof of this theorem that a curious reader will find interesting.

To prove this theorem, it is important first of all to show that the factorization exists. In other words, we have to prove that any given integer n greater than or equal to 2 is prime itself or it can be written as the product of prime numbers. Once this is done, we can then prove the uniqueness of the prime decomposition.

Starting with the first step, we will prove, using the proof by induction method* (precisely the strong induction), that if n is an integer such that n > 1, then n is either a prime number or it is the product of prime numbers.

---

*The proof by induction

It is a method used in mathematics to prove that a statement involving a natural number n holds for any value of n by proving that it holds for an **initial case**, also called the **base case**(usually, by substituting the value of n with 0 or 1 or another integer depending on the exact statement to be proved). The next step is called the **induction step** and it consists of proving that if this statement holds an arbitrary number n, then it must hold also for n+1.

Technical, usually, when writing a proof that uses induction, it is recommended to write a neat conclusion of what you have just proved.

Mathematical Induction is one of the most interesting and widely-used methods of prooving statements in number theory.

A simple and intuitive way of illustrating the nature of induction is by thinking of the Domino effect. Proving the Base case can be seen as toppling by hand the first domino in the sequence. The step of induction can be interpreted as: if the nth domino falls, the next piece (n+1) will, too.

36 The fundamental Theorem of Arithmetic

> *Mathematical induction proves that we can climb as high as we like on a ladder, by proving that we can climb onto the bottom rung (the **basis**) and that from each rung we can climb up to the next one (the **step**).* [4]
>
> Strong induction is a variant of induction in which we assume that the statement holds for all values preceding n, whereas in the simple induction we assume that the statement holds only for n.

By strong induction, we will assume that every integer k strictly greater than 1 and less than n is either a prime or a product of prime numbers.

If n is prime, then we have nothing to prove. However, if n is not prime, i.e. is composite, then, by definition, n must be a product of two positive integers a and b that are strictly greater than 1 and less than n, ($1 < b \leq a < n$ or $1 < a \leq b < n$). Since a and b are less than n, then, by our induction hypothesis, both a and b are the product of prime numbers, i.e.,

$$a = p_1^{\alpha_1} \times p_2^{\alpha_2} \times ... \times p_k^{\alpha_k} \text{ and } b = q_1^{\beta_1} \times q_2^{\beta_2} \times ... \times q_k^{\beta_k},$$

where p_i and q_i are distinct prime numbers such that $p_1 \leq p_2 \leq ... \leq p_k, q_1 \leq q_2 \leq ... \leq q_k$, $1 \leq i \leq k$, and $\alpha_i, \beta_i \geq 1$. Thus,

$$n = a \times b = p_1^{\alpha_1} \times p_2^{\alpha_2} \times ... \times p_k^{\alpha_k} \times q_1^{\beta_1} \times q_2^{\beta_2} \times ... \times q_k^{\beta_k}$$

is a product of prime numbers.

Therefore, we conclude that for any given integer $n > 1$, n is either a prime or it is a product of prime numbers.

The second step is proving the uniqueness of the prime factorization.

To do so, we will use the proof by contradiction method.

By way of contradiction, suppose that the factorization of n is not unique. Thus,

$$n = p_1 \times p_2 \times ... \times p_k = q_1 \times q_2 \times ... \times q_l,$$

where $l, k \geq 2$, and p_i and q_j are prime numbers such that at least one of p_i is different from all the q_j, otherwise, the factorization will be the same.

The fundamental Theorem of Arithmetic

(For the sake of simplicity and without loss of generality, we take the exponent of all the prime factors of n as 1, i.e., n is a square-free number, thus the p_i are distinct primes, and the same goes for q_j).

Based on $p_1 \times p_2 \times ... \times p_k = q_1 \times q_2 \times ... \times q_l$, we can deduce that p_1 divides $q_1 \times q_2 \times ... \times q_l$. By Euclid's Lemma*, p_1 divides one of $q_1, q_2, ..., q_l$.

Without loss of generality, let's assume that p_1 divides q_1. Since p_1 and q_1 are both primes, $p_1 = q_1$. Canceling out p_1 and q_1 from both sides, we get

$$p_2 \times p_3 ... \times p_k = q_2 \times q_3 \times ... \times q_l.$$

The same reasoning can be applied to prove that every p_i is equal to one of the q_j. But this is absurd since we supposed that the two factorizations of n are distinct. Therefore, our hypothesis is false, which is to say that the prime factorization of n is unique.

■

Euclid's Lemma:

Euclid's Lemma states that if a prime number p divides a product of two numbers a and b, i.e., divides a × b, then p must divide at least one of those two integers. In other words, p divides a, or p divides b, or p divides both a and b.

Note:

Writing n as the product of prime numbers given in increasing order of magnitude is called **the standard factorization of n**.

Sieving for Primes

How can we generate a list of prime numbers?

Well, one of the simplest and most famous methods is sieving. It is used to find prime numbers up to a given limit.

According to the **Dictionary.com**, the linguistic definition of "to sieve" is:

- to put or force through a sieve;
- to sift: to separate and retain the coarse parts of (flour, ashes, etc.) with a sieve. [1+2]

As the definition states, sieving means separating something from something else. More generally, sieving is not only used with the set of prime numbers but also it can be used with an arbitrary set of numbers. There are many methods of sieving and they form sieve theory, an interesting area in number theory.

Sieve of Eatosthenes

One of the most popular algorithms of sieving, and it is also among the first to be discovered, is called the sieve of Eratosthenes, suggested by **Eratosthenes of Cyrene** (276-194 BC) who was a Greek mathematician, astronomer, and poet. He is best known for his estimation of the circumference of the earth and the distances to the Moon and sun. [3]

Sieving for Primes

The sieve of Eratosthenes is one of the easiest and fastest ways of generating a list of ordered prime numbers.

It works in the following way:

First, create a list of positive integers starting from 2.

Let's say the first 99 integers. 2 is a prime number. Thus, all the multiples of 2, i.e. all even numbers (2, 6, 8, 10, 12, ...), are composite. So, we strike them out.

Figure 1:
Eratosthenes of Cyrene
(276-194 BC)

	2	3	4	5	6	7	8	9	10
11	12	13	14	15	16	17	18	19	20
21	22	23	24	25	26	27	28	29	30
31	32	33	34	35	36	37	38	39	40
41	42	43	44	45	46	47	48	49	50
51	52	53	54	55	56	57	58	59	60
61	62	63	64	65	66	67	68	69	70
71	72	73	74	75	76	77	78	79	80
81	82	83	84	85	86	87	88	89	90
91	92	93	94	95	96	97	98	99	100

Figure 2:
Sieve of Eratosthenes: even numbers crossed out

The next smallest uncrossed number is 3. It is also a prime number. So, all the multiples of 3 up to the given limit (6, 9, 12, 15, ...) are crossed out.

40 Sieving for Primes

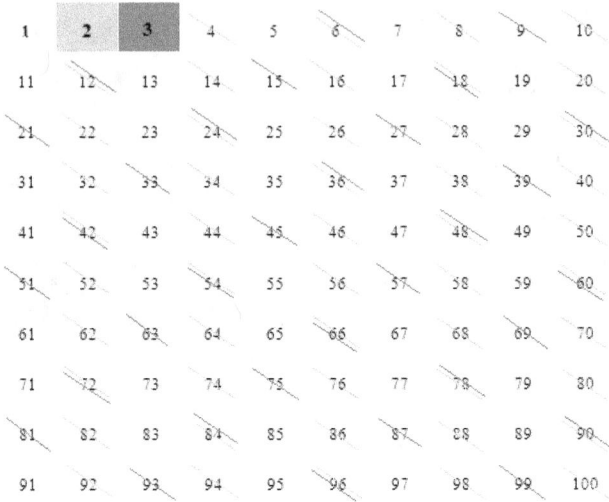

Figure 3:
Sieve of Eratosthenes: even numbers and multiples of 3 crossed out
crossed out

Next, we have 5. Multiples of 5 (10, 15, 20, 25, 30,...), OUT! ... We repeat this procedure until all composite numbers in the list are crossed out, and we will be left only with prime numbers.

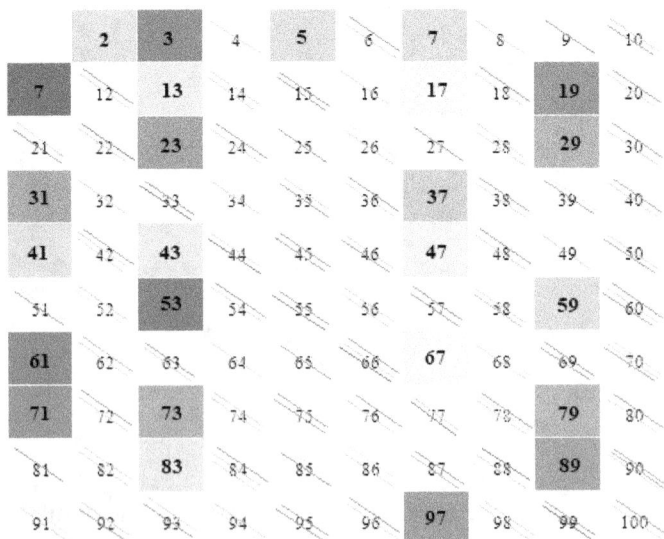

Figure 4:
Sieve of Eratosthenes: all composite numbers are crossed out and only prime numbers are left

Sieving for Primes 41

Note that, for example, when you strike out the multiples of 3, some of them (6, 12, 18, …) will be already crossed out since they are also multiples of 2. Hence, many numbers will be crossed out many times. [4]

You may like the following ditty about the sieve of Eratosthenes from the American writer **Frederik Pohl**'s 1960 book *Drunkard's Walk*:

> *Strike the Twos and strike the Threes,*
>
> *The Sieve of Eratosthenes!*
>
> *When the multiples sublime,*
>
> *The numbers that are left, are prime.* [5]

We can optimize more this method.

If the given limit is N, then we have to check only the numbers less than $\sqrt{N}$.

This is easy to prove. Let K be a positive integer. If k is not prime, then it is composite and can be represented as the product of two positive integers a and b (not necessarily primes). If $a > \sqrt{k}$ and $b > \sqrt{k}$, then $ab > K$ and if $a < \sqrt{k}$ and $b < \sqrt{k}$, then $ab < K$. However, this is absurd since $k = ab$. Then, we can deduce that one of a and b must be less than $\sqrt{k}$.

∎

This is a useful trick when checking the primeness of a given number. Thus, when using trial division*, you don't have to check all the primes less than the number in question. The primes less than its square root will be enough.

> *Trial division
>
> This is one of the oldest and the simplest methods to check the primeness of a given number. To do so, first, we have to divide the number in question by 2, if it has no remainder, then it is composite. If it has, then we divide it by 3. If it has no remainder, then it is composite. We repeat this procedure for all prime numbers less than $\sqrt{k}$ for a reason we explained earlier. [6]

42 Sieving for Primes

For example, let's consider the number 829 153. It is important to calculate the square root of 829153 before we start. ($\sqrt{829153} \approx 910.578 ...$). It is obviously not divisible by 2. Dividing this number by 3 gives 1 as a remainder. Division by 5, 7, and 11 gives a remainder of 3, 3, and 6, respectively. However, upon division by 13, we get 0 as a remainder. Therefore, 829 153 is composite. You can easily determine its other factor which is 63 781.

As another example, let's consider the number 103. We calculate first of all its square root, that is $\sqrt{103} \approx 10.149$ .... Thus, we have to check only prime numbers less than 10. 103 is divisible neither by 2, by 3, nor by 5. When divided by 7, 103 gives a remainder of 5. Therefore, we can conclude that 103 is prime.

However, this method is not unique. Indeed, there exist many others.

Sieve of Euler

The famous Swiss mathematician, **Leonhard Euler**, invented a sieving algorithm that was called after him, Euler's sieve. While in the sieve of Eratosthenes some (composite) numbers can be crossed out multiple times, in the Euler's sieve, every composite number is deleted only once.

In Euler's sieve, we start also with an ordered list of integers ranging from 2 to a given number n. In each step, we take the first element of the list (which is always a prime number) and multiply it by all the numbers in that list.

Figure 5:
Leonhard Euler (1707 - 1783)

The resulting numbers from this multiplication, as well as the first element, will be crossed out from the original list. Next, you have to repeat this procedure until the original list becomes empty. Please note that you have to make a second list where you save the first element of the original list in each step. At the end, the second list will contain all the prime numbers less than n. [7]

44 Sieving for Primes

Example

For example, let's take, as the original list, numbers ranging from 2 to 30, and let's call this list A. We will create another list B that will be our resulting list. Initially, B is empty.

A | 2, 3, 4, 5, 6, 7, 8, 9, 10, 11, 12, 13, 14, 15, 16, 17, 18, 19, 20, 21, 22, 23, 24, 25, 26, 27, 28, 29, 30

The first number in this list is 2. We will add 2 to B. Then, we will multiply 2 by all the elements of A and, for the sake of clarity, we will save the result in a temporary list C.

A | 2, 3, 4, 5, 6, 7, 8, 9, 10, 11, 12, 13, 14, 15, 16, 17, 18, 19, 20, 21, 22, 23, 24, 25, 26, 27, 28, 29, 30

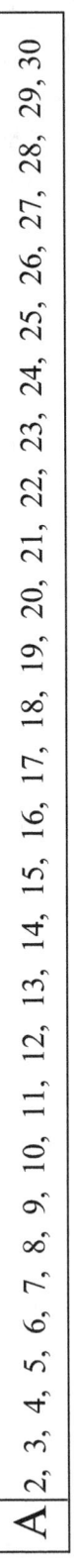

C | 4, 6, 8, 10, 12, 14, 16, 18, 20, 22, 24, 26, 28, 30, 32, 34, 36, 38, 40, 42, 44, 46, 48, 50, 52, 54, 56, 58, 60

B

Then, we will "subtract" C from A, i.e., in the next step, A will contain the intersection between A in the current step and C. In other words, we will delete from A its elements that are in common with C. Also, the first element of A, which is in this case 2, will be eliminated.

Thus, the A will be

$$A \mid 3, 5, 7, 9, 11, 13, 15, 17, 19, 21, 23, 25, 27, 29$$

Note: You may have noticed that, at a certain point (32), the elements in C become greater than the maximum number in A. Thus, not including them makes no difference. In other words, when we multiply the first element of A by one of its other elements and the result is greater than the maximum element of A, we can stop this task, i.e., we have to calculate the elements of C that are within the range of the original list A.

In this case, C will be

$$C \mid 4, 6, 8, 10, 12, 14, 16, 18, 20, 22, 24, 26, 28, 30$$

The next step will be similar to the previous one

46 Sieving for Primes

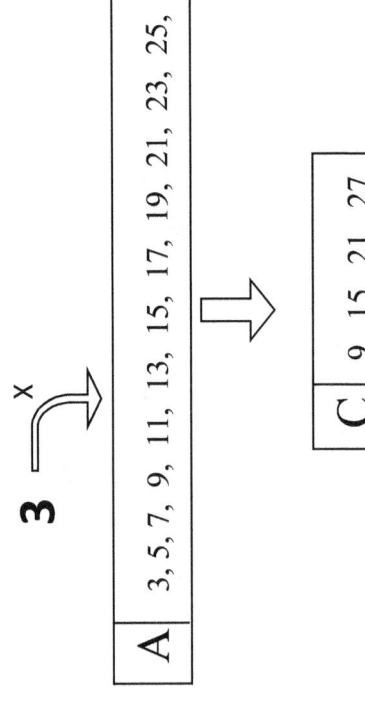

Subtracting C from A and the latter's first element, we will get

| A | 5, 7, 11, 13, 17, 19, 23, 25, 29 |

Repeating the same process for 5, we get

| A | 7, 11, 13, 17, 19, 23, 29 | | B | 2, 3, 5 |

Then,

| A | 11, 13, 17, 19, 23, 29 |

| B | 2, 3, 5, 7 |

We repeat this procedure until A is empty. At that point, B will contain all the prime numbers less than 30.

| A | |

| B | 2, 3, 5, 7, 11, 13, 17, 19, 23, 29 |

48   Sieving for Primes

Here are the steps of the sieve of Euler in an "algorithmic" way:

1. Make a first list containing the integers from 2 to n and a second list that will contain the prime numbers (the resulting list).
2. Extract the first element from the first list (original list) and insert it in the resulting list.
3. Make another list in which each element from the original list (including the first) is multiplied by the element extracted in step 2.
4. Subtract the new list from the original list.
5. Repeat the 2nd, 3rd, 4th, and 5th step until the original list is empty.
6. The algorithm is finished and the resulting list will contain all the prime numbers less than n.

Optimization of the sieve of Euler

In addition to the optimization concerning calculating the elements of the temporary list (C in the last example), there are other tricks to improve this algorithm.

One way is by including only odd numbers in the original list since all prime numbers are odd except 2. The 2 will be dealt with separately.

Another way is by checking only numbers from the original list up to $\sqrt{n}$ (as explained earlier).

* * *

Another method of sieving is called the sieve of Atkin. This is a modern algorithm that instead of crossing out the multiples of prime numbers, it eliminates multiples of squares of prime numbers. It was introduced in 2003 by the American-German mathematician **Daniel Julius Bernstein** and the British mathematician **Arthur Oliver Lonsdale Atkin** (usually referred to as **A. O. L. Atkin**) [8]

There exist also the sieve of Sundaram, discovered by an obscure Indian mathematician called **S.P. Sundaram** in the 1930s. [9]

Chapter - 7

The Prime-counting funtion π

Using sieving methods (Sieve of Eratosthenes, Sieve of Euler, ...), we can determine all the prime numbers less than or equal to a given positive integer n. However, sometimes, we need to know only how many prime numbers are less than a certain number. For that reason, mathematicians defined a new function called the prime-counting function and often denoted by $\pi(x)$, where x is a positive real number*. (it has no connection with the geometric constant $\pi \approx 3.14$ ...)

$\pi(x)$ is the number of prime numbers less than or equal to x, succinctly,
$\pi(x) = |\{p \text{ a prime number}; \ p \leq x\}|$

where "| |" denotes the cardinality of a set, i.e., the number of its elements.

*Note that the prime-counting function $\pi(x)$ is defined over the set of positive real numbers. This may be counter-intuitive for the first time. However, it is important to mention that since the 19th century, mathematicians started to develop and use ''analytic'' methods, such as calculus, to study prime numbers. Using calculus to study the purest form of numbers may be shocking for the first time. Nevertheless, this new approach was fruitful and led to the discovery of many important results in number theory.   [1+2]

The Prime-counting function π

For example,

π(10) = 4 since there are 4 prime numbers less than or equal to 10 and they are 2, 3, 5, and 7.

π(2) = 1 since there is only one prime number less than or equal to 2 and it is obviously 2.

π(1) = 0 because there are no prime numbers less than or equal to 1.

The first few values of $\pi(10^n)$, where n is a natural number are:

n	10^n	$\pi(10^n)$
1	10	4
2	100	25
3	1,000	168
4	10,000	1,229
5	100,000	9,592
6	1,000,000	78,498
7	10,000,000	664,579
8	100,000,000	5,761,455
9	1,000,000,000	50,847,534
10	10,000,000,000	455,052,511
11	100,000,000,000	4,118,054,813
12	1,000,000,000,000	37,607,912,018
13	10,000,000,000,000	346,065,536,839

[1+3]

The sequence of terms of the $\pi(10^n)$, where n is an integer, can be found on the Online Encyclopedia for Integer Sequences' website (Sequence A006880 on oeis.org) [4]

> **Note:**
>
> The OEIS stands for the **Online Encyclopedia for Integer Sequences**, also called **Sloane's Encyclopedia**. As its name may explaine, it is an online database that contains more than 336 201 sequences (August 2020). The encyclopedia was created in 1964 by the British-American mathematician Niel Sloane. However, the online version of it was launched in 1996.
>
> It will be a great idea to new amazing sequences on their website: www.oeis.org
>
> As of the AXXXXXX, it is the id number that is unique for every sequence on OEIS's website.

The following is the graph of the function $\pi(x)$.

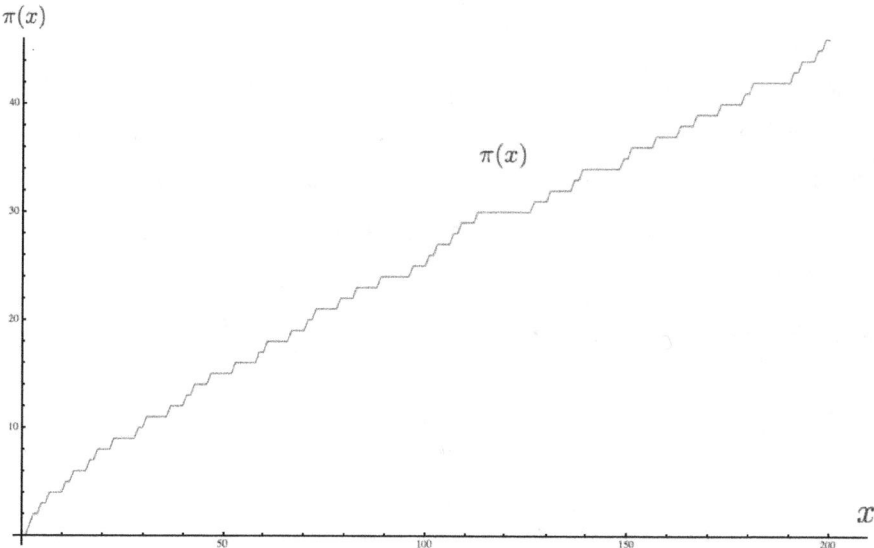

Figure 1: Graphical representation of the prime-counting function π(x)

52 The Prime-counting function π

Based on the graph, you may have noticed that this function is strictly increasing (for x > 0) and it is indeed since the number of prime numbers is infinite. The last statement can be expressed concisely using the notion of the limit as [2]

$$\lim_{x \to \infty} \pi(x) = \infty$$

The prime-counting function is extremely important in the study of prime numbers. In fact, one of the famous theorems in number theory involving this function is the Prime Number Theorem (PNT). It is an important resultsince it describes the increasing behavior of the function $\pi(x)$.

A slightly different version of the PNT was first stated in 1798 by French mathematician **Adrien-Marie Legendre** (1752 - 1833) in his book *Essai sur la Théorie des Nombres*.[5] However, it was first proved in 1896, independently, by Frech mathematician **Jacques Hadamard** (1865 – 1963) and Belgian mathematician **Charles Jean de la Vallée Poussin** (1866 –1962). Their work was based on a revolutionary paper written by the famous mathematician **Bernhard Riemann** and concerning the distribution of prime numbers. [6]

PNT states that **as n gets bigger, i.e., as n tends to infinity, $\pi(x)$ asymptotically approches** $\frac{x}{\ln(x)}$, where ln(x) is the natural logarithm of x. The former statement is just a fancy way of saying that when we graph the function $\pi(x)$ and the function $\frac{x}{\ln(x)}$, we will notice that they grow approximately at the same rate. (Think of it as they are roughly the same for larger values of x).

This can be expressed differently, using the notion of the limit, as

$$\lim_{x \to \infty} \frac{\pi(x)}{\frac{x}{\ln(x)}} = 1$$

This may also be denoted by $\pi(x) \sim \frac{x}{\ln(x)}$, where "~ " is the asymptotic notation. [2]

The following graph may help clarify this concept.

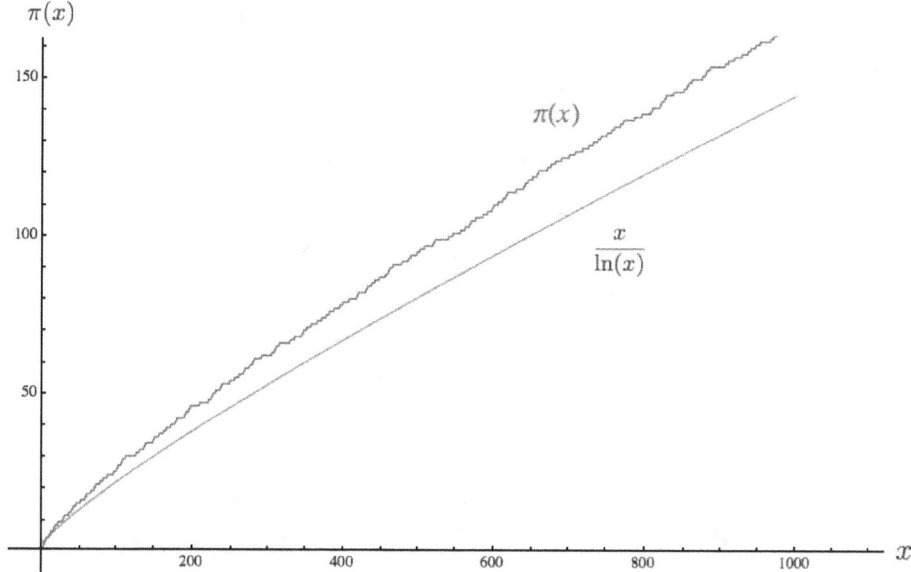

Figure 2: Graphical representation of π(x) and x/ln(x)

PNT is useful in determining the average gap between consecutive prime numbers less than a given number n, which is approximately ln(n).[7] For example, if n is around 100, then ln(100) ≈ 4.6 . Hence, under 100, we can expect to see a prime number approximately every 4 or 5 consecutive numbers. However, this becomes more accurate as the given limit n tends to infinity.

A better approximation of $\pi(x)$ was discovered by German mathematician **Carl Friedrich Gauss.** He showed that $\pi(x)$ is roughly equivalent to the offset Logarithmic integral $Li(x)$ defined as [9]

$$Li(x) = \int_2^x \frac{2}{\ln(t)} dt$$

Without getting into many details, this can be illustrated using the following graph.

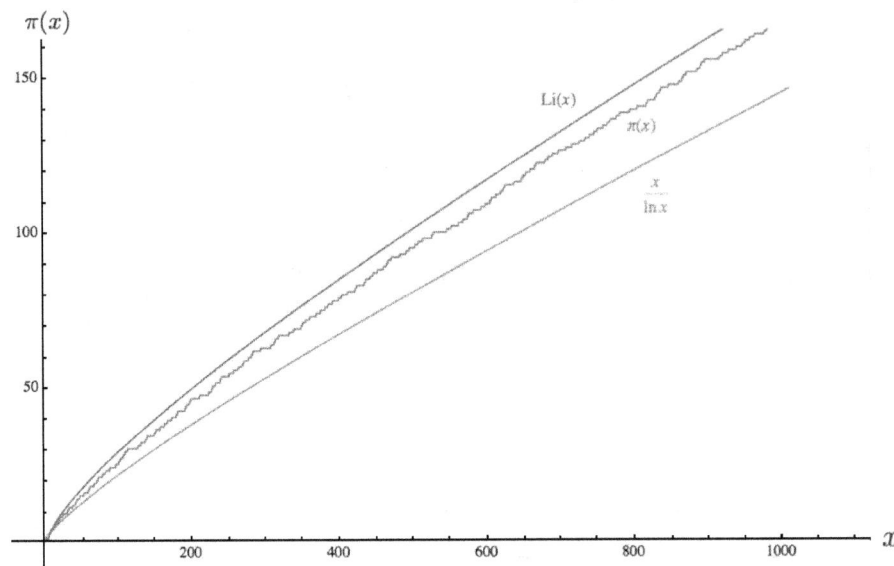

Figure 3: Graphical representation of π(x), x/ln(x), and Li(x)

In general, Li(x) is a better approximation to $\pi(x)$.

We can even observe this numerically in the following table that compares the error (to the nearest integer) of the Li(x)'s and $\frac{x}{\ln(x)}$'s estimation of $\pi(x)$. [10]

x	$\pi(x)$	$\frac{x}{\ln(x)} - \pi(x)$	$Li(x) - \pi(x)$
10	4	0	2
10^2	25	-3	5
10^3	168	-23	10
10^4	1,229	-143	17
10^5	9,592	-906	38
10^6	78,498	-6,116	130
10^7	664,579	-44,158	339
10^8	5,761,455	-332,774	754
10^9	50,847,534	-2,592,592	1,701
10^{10}	455,052,511	-20,758,029	3,104
10^{11}	4,118,054,813	-169,923,159	11,588
10^{12}	37,607,912,018	-1,416,705,183	38,263
10^{13}	346,065,536,839	-11,992,858,452	108,971

The Prime-counting function π

There exists another way of calculating the prime-counting function $\pi(x)$ that is simple, however, using it is a time-consuming task, especially when x gets bigger and bigger.

Discovered by **Legendre**, it states that:

$$\pi(x) = -1 + \pi\left(\sqrt{x}\right) + \lfloor x \rfloor - \sum_{p_i \leq \sqrt{x}} \left\lfloor \frac{x}{p_i} \right\rfloor + \sum_{p_i < p_j \leq \sqrt{x}} \left\lfloor \frac{x}{p_i\, p_j} \right\rfloor -$$

$$\sum_{p_i < p_j < p_k \leq \sqrt{x}} \left\lfloor \frac{x}{p_i\, p_j\, p_k} \right\rfloor + \sum_{p_i < p_j < p_k < p_l \leq \sqrt{x}} \left\lfloor \frac{x}{p_i\, p_j\, p_k\, p_l} \right\rfloor - \dots$$

> **Note:**
>
> $\lfloor x \rfloor$ denotes the integer part of x.
>
> This function, called the floor function, is defined as:
>
> If $n \leq x < n+1$ where x is a real number and n is an integer, then $\lfloor x \rfloor = n$.
>
> In other words, $\lfloor x \rfloor$ is the biggest integer smaller than or equal to x.
>
> For example, $\lfloor 3.14 \rfloor = 3$, $\lfloor 4.71 \rfloor = 4$, $\lfloor -7.11 \rfloor = -8$, $\lfloor 5 \rfloor = 5$, ...

This formula may seem complicated. However, if you think about it for a while, you will know how easy it is.

Let's think about the integers of the interval [1, x] (numbers less than n and greater than 1). Those integers can be classified into prime and composite numbers.

Based on the former arguments, we get the following "pseudo-formula":

$$\begin{pmatrix} \text{the number of all} \\ \text{positive integers} \\ \text{up to } x \end{pmatrix} - 1 = \begin{pmatrix} \text{number of prime} \\ \text{numbers up to } x \end{pmatrix} + \begin{pmatrix} \text{number of} \\ \text{composite} \\ \text{numbers up to } x \end{pmatrix}$$

If you are wondering about the -1, we subtracted it because the number of all the positive integers contains 0, but we don't want to count it. We just need the number of integers from 1 to x.

The simplified formula above can be written differently as

$$\begin{pmatrix} \text{number of prime} \\ \text{numbers up to } x \end{pmatrix} + 1 = \begin{pmatrix} \text{the number of all} \\ \text{positive integers} \\ \text{up to } x \end{pmatrix} - \begin{pmatrix} \text{number of} \\ \text{composite} \\ \text{numbers up to } x \end{pmatrix}$$

Ley's go back to our initial formula.

The number of all the integers less than or equal to x is known and it is simply $\lfloor x \rfloor$. So, all we have to do to get the number of primes is to calculate the number of composite numbers.

Let p be a prime number. Then, $\lfloor \frac{x}{p} \rfloor$ will be the number of positive integers in the interval $[1, x]$ and divisible by p. For example, if $x = 10$ and $p = 2$, then the number of integers less than or equal to 10 and that are divisible by 2 is $\lfloor \frac{10}{2} \rfloor = 5$. We can determine them easily, indeed, they are 2, 4, 6, 8, and 10. Let $p_i = 3$. The number of integers in $[1, 10]$ that are divisible by 3 is $\lfloor \frac{10}{3} \rfloor = 3$. Those numbers are 3, 6, and 9.

Hence, $\sum_{p_i \leq \sqrt{x}} \lfloor \frac{x}{p_i} \rfloor$ is the sum of all the integers in $[1, x]$ that are divisible by the prime numbers p_i (the p_i are less than $\sqrt{x}$ for a reason we discussed earlier). So we subtract it. But, in this sum, we have also calculated the prime numbers p_i (that are less than or equal to $\sqrt{x}$) because they are multiples of themselves. Therefore, we have to add $\pi(\sqrt{x})$ (which is the number of prime numbers up to $\sqrt{x}$).

However, in this sum of composite numbers that we calculated, some composite numbers, precisely those that are divisible by two prime numbers, are counted twice. For example, if a number is multiple of, say 2 and 3, it will be counted first as a multiple of 2, and then will be counted for the second time as a multiple of 3. So, we must correct our total by adding the

The Prime-counting function π   57

sum of integers that are divisible by two primes ($\leq \sqrt{x}$), i.e.,
$\sum_{p_i < p_j \leq \sqrt{x}} \left\lfloor \frac{x}{p_i p_j} \right\rfloor$.

However, by adding this new term, composite numbers that are multiples of three prime numbers won't be counted. Thus, we will have to subtract them, i.e., subtract $\sum_{p_i < p_j < p_k \leq \sqrt{x}} \left\lfloor \frac{x}{p_i p_j p_k} \right\rfloor$. And so on…

■ [11]

For example, let's use this method to calculate $\pi(100)$.

Obviously, the primes less than or equal to $\sqrt{100} = 10$ are 2, 3, 5, and 7.

$$\begin{aligned}\pi(100) = &-1 + \pi(10) + \lfloor 100 \rfloor - \left\lfloor \frac{100}{2} \right\rfloor - \left\lfloor \frac{100}{3} \right\rfloor - \left\lfloor \frac{100}{5} \right\rfloor - \left\lfloor \frac{100}{7} \right\rfloor \\ &+ \left\lfloor \frac{100}{2 \times 3} \right\rfloor + \left\lfloor \frac{100}{2 \times 5} \right\rfloor + \left\lfloor \frac{100}{2 \times 7} \right\rfloor + \left\lfloor \frac{100}{3 \times 5} \right\rfloor + \left\lfloor \frac{100}{3 \times 7} \right\rfloor \\ &+ \left\lfloor \frac{100}{5 \times 7} \right\rfloor - \left\lfloor \frac{100}{2 \times 3 \times 5} \right\rfloor - \left\lfloor \frac{100}{2 \times 3 \times 7} \right\rfloor - \left\lfloor \frac{100}{2 \times 5 \times 7} \right\rfloor \\ &- \left\lfloor \frac{100}{3 \times 5 \times 7} \right\rfloor + \left\lfloor \frac{100}{2 \times 3 \times 5 \times 7} \right\rfloor \\ = &-1 + 4 + 100 - 50 - 33 - 20 - 14 + 16 + 10 + 7 + 6 + 4 \\ &+ 2 - 3 - 2 - 1 - 0 + 0 \\ = &\; 25.\end{aligned}$$

I am sure you will be able to verify this result on your own.

(You can check it manually using the sieve of Eratosthenes, or you can just verify it in the previous chapter).

Chapter 8

Euler's totient function

Euler's phi (φ) function, also known as Euler's totient function, of a positive integer n, is equal to the number of integers that are greater than or equal to 1 and smaller than or equal to n, and which are relatively prime with n. Those numbers are called the totatives. [1]

Examples:

$\varphi(2) = 1$ since there is only one number less than 2 and bigger or equal to 1 that is relatively prime with 2, and it is 1.

$\varphi(4) = 2$ because 1 and 3 are the only integers bigger than or equal to 1 and less than 4, and that does not share any common prime factor with 4.

$\varphi(11) = 2$, $\varphi(24) = 8$ (the totatives are: 1, 5, 7, 11, 13, 17, 19, and 23), ...

$\varphi(1) = 1$ because there is only the number 1 in the range from 1 to 1 such that gcd(1, 1) =1.

The first values of this function are

Integer n	1	2	3	4	5	6	7	8	9	10	11	12	13	14	15	16
$\varphi(n)$	1	1	2	2	4	2	6	4	6	4	10	4	12	6	8	8

You can find more on the Online Encyclopedia of Integer Sequences (OEIS)'s website. (Sequence A000010 in OEIS)

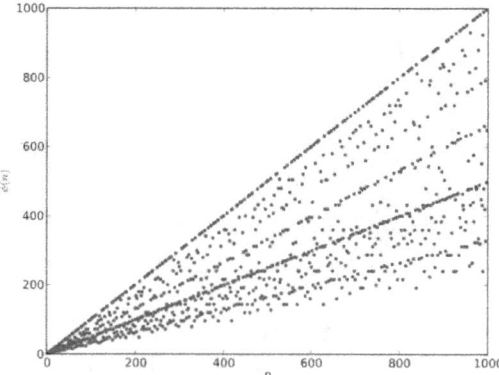

Figure 1: The first thousand values of φ(n).

The points on the top line represent φ(p) when p is a prime number, which is p − 1.

Some properties of φ(n)

A good observation is that if p is a prime number, then

$$\varphi(p) = p - 1,$$

since there exist p - 1 integers k such that k ≤ p and k ≥ 1, that don't share any common factor with p. (This follows from the definition of a prime number).

Another property of Euler's phi function is that it is a multiplicative function. This is just a fancy way of saying that $\varphi(m \times n) = \varphi(m) \times \varphi(n)$, where m and n are two positive co-prime integers. [2]

In general, If m and n are two integers (not necessarily co-prime), then

$$\varphi(m \times n) = \varphi(m) \times \varphi(n) \times \frac{d}{\varphi(d)},$$

where d = gcd(m, n).

For example,

φ (60) = φ (22) × φ (3) × φ(5) = (4 − 2)(3 − 1)(5 − 1) = 16.

It follows that if p is a prime number, then

$$\varphi(p^n) = p^{n-1} \times (p - 1)$$

60 Euler's Totient function

The proof of the former statement is simple.

Let p be a prime number and k a positive integer.

Keep in mind that we are searching for numbers less than p that are relatively prime with p^k.

Let a be an integer in the range from 1 to p. For the gcd of a and p^k to be 1, a must not be a multiple of p, otherwise, the gcd will be itself a multiple of p.

The multiples of p that are less than or equal to p^k are $1 \times p$, $2 \times p$, $3 \times p$, $4 \times p$, ..., $p^{k-1} \times p$. Thus, there are p^{k-1} integer satisfying this condition.

Based on the last two results, we can deduce that there are $p^k - p^{k-1}$ integer between 1 and p^k (inclusive) that are not multiples of p, hence, they are relatively prime with p^k.

So, $\varphi(p^k) = p^k - p^{k-1} = p^{n-1} \times (p-1)$. ■[3]

Swiss mathematician **Leonhard Euler** (1707 – 1783) suggested a straightforward formula that can be useful when calculating $\varphi(n)$, where n is an integer. [3]

He proved that $\varphi(n)$ is not but n multiplied by the product of $1 - \frac{1}{p_i}$, where p_i are all the distinct prime numbers that divide n.

We can write the last statement succinctly as:

$$\varphi(n) = n \times \prod_{p\,;\,p|n} \left(1 - \frac{1}{p}\right)$$

where Π is the pi product notation(It is a capital Greek letter that is used to represent a product of a bunch of terms) and " | " means "divides ". p|n is the same way of the saying p divides n.

In other words, if $n = p_1^{\alpha_1} \times p_2^{\alpha_2} \times p_3^{\alpha_3} \times p_4^{\alpha_4} \times ... \times p_k^{\alpha_k}$, where p_i are distinct prime numbers such that $p_1 \leq p_2 \leq p_3 \leq ... \leq p_k$, $1 \leq i \leq k$, and $\alpha_i \geq 1$ then

Euler's Totient function

$$\varphi(n) = n \times \left(1 - \frac{1}{p_1}\right)\left(1 - \frac{1}{p_2}\right)\left(1 - \frac{1}{p_3}\right) \times \ldots \times \left(1 - \frac{1}{p_k}\right)$$

From this formula, we can deduce that if $n = p^k$, then

$$\varphi(n) = \varphi(p^k) = p^k \times \left(1 - \frac{1}{p}\right) = p^{k-1} \times (p - 1).$$

Even though this formula may seem a bit complicated, it has a simple proof.

Let $n = p_1^{\alpha_1} \times p_2^{\alpha_2} \times p_3^{\alpha_3} \times p_4^{\alpha_4} \times \ldots \times p_k^{\alpha_k}$, where p_i are distinct prime numbers such that $p_1 \leq p_2 \leq p_3 \leq \ldots \leq p_k$, $1 \leq i \leq k$, and $\alpha_i \geq 1$. Then,

$$\varphi(n) = \varphi(p_1^{\alpha_1} \times p_2^{\alpha_2} \times p_3^{\alpha_3} \times p_4^{\alpha_4} \times \ldots \times p_k^{\alpha_k})$$

Since Euler's totient function is multiplicative, we get that

$$\varphi(n) = \varphi(p_1^{\alpha_1}) \times \varphi(p_2^{\alpha_2}) \times \varphi(p_3^{\alpha_3}) \times \varphi(p_4^{\alpha_4}) \times \ldots \times \varphi(p_k^{\alpha_k})$$

We know that if p is a prime number and k is a positive integer, then

$$\varphi(p^\alpha) = p^{\alpha-1} \times (p - 1) = p^\alpha \times \left(1 - \frac{1}{p}\right)$$

Thus,

$$\varphi(n) = p_1^{\alpha_1} \times \left(1 - \frac{1}{p_1}\right) \times p_2^{\alpha_2} \times \left(1 - \frac{1}{p_2}\right) \ldots \times p_k^{\alpha_k} \times \left(1 - \frac{1}{p_k}\right)$$

$$= p_1^{\alpha_1} \times p_2^{\alpha_2} \times \ldots \times p_k^{\alpha_k} \times \left(1 - \frac{1}{p_1}\right) \times \left(1 - \frac{1}{p_2}\right) \times \ldots \times \left(1 - \frac{1}{p_k}\right)$$

$$= n \times \prod_{p\,;\,p|n} \left(1 - \frac{1}{p}\right)$$

∎

Calculating phi of any positive integer will be easier using this function since only the prime decomposition is needed.

Euler, who introduced this function for the first time in 1763, used it in a theorem that is considered, nowadays, one of the most important results in number theory. Speaking of which, it is called Euler's Theorem. [4]

This theorem states that **if a and n are co-prime positive integers, then**

$$a^{\varphi(n)} \equiv 1 \pmod{n}$$

Euler's Totient function

A special case of this theorem is when n is prime. It becomes

$$a^{\varphi(n)} \equiv a^{n-1} \equiv 1 \pmod{n}.$$

This is known as Fermat's Little Theorem. (This is covered in details in the chapter dedicated to Fermat's work)

One of the modern applications of this theorem is the RSA cryptosystem that we discussed in the third chapter.

In addition to the ones stated earlier, this function has many other properties.

In fact, if a and b are two positive integers such that a divides b, then $\varphi(a)$ divides $\varphi(b)$.

Also, $\varphi[\gcd(a,b)] \times \varphi[\text{lcm}(a,b)] = \varphi(a) \times \varphi(b)$. Doesn't this remind you of $\gcd(a,b) \times \text{lcm}(a,b) = a \times b$?

Another interesting property of this function is that $\varphi(n)$ is always even when n ≥ 3. Proving this will be a challenge for the reader. (**Hint:** the proof is easy if you use one of the formulas we proved earlier in this chapter)

Another interesting property is that if n is a positive integer that has k distinct odd prime factors (prime factors different from 2), then 2^k divides $\varphi(n)$. To prove the last statement, we will use the one before.

Let $n = p_1^{\alpha_1} \times p_2^{\alpha_2} \times p_3^{\alpha_3} \times p_4^{\alpha_4} \times ... \times p_k^{\alpha_k}$, where p_i are distinct prime numbers such that $p_1 \leq p_2 \leq p_3 \leq ... \leq p_k$, $1 \leq i \leq k$, and $\alpha_i \geq 1$. We know that $\varphi(p^\alpha) = p^{\alpha-1} \times (p-1) = 2q$, where q is some positive integer (we don't need to care about this number, what matters is just the fact that $\varphi(p^\alpha)$ is even)
Therefore,
$\varphi(n) = 2q_1 \times 2q_2 \times 2q_3 \times ... \times 2q_k = 2^k(q_1 \times q_2 \times q_3 \times ... \times q_k)$.
Thus, 2^k divides $\varphi(n)$.

Perfect totient numbers

A number m is said to be a totient number if there exists at least a positive integer n such that $\varphi(n) = m$. A non-totient number m, however, is a number for which $\varphi(n) \neq m$ for any positive integer n. A trivial example of non-totient numbers is the odd numbers greater than 2. (This follows from a property of Euler's totient function that assures the evenness of $\varphi(n)$ for $n \geq 3$).

There is a special kind of totient numbers, and those numbers are called the perfect totient numbers.

A number is said to be a perfect totient number if it equals the sum of its iterated totients. In other words, we apply Euler's phi function to a number n, re-apply it to the first result, re-re-apply it to the second result, and so on and so fourth until we reach one, and finally we add all those results. If the sum is n, then n is called a perfect totient number. [5]

For example, let n = 15.

$\varphi(15) = 8 \Rightarrow \varphi(8) = 4 \Rightarrow \varphi(4) = 2 \Rightarrow \varphi(2) = 1$

$1 + 2 + 4 + 8 = 15$. Thus, 15 is a perfect totient number.

However, 24 is not a perfect totient number because:

$\varphi(24) = 8 \Rightarrow \varphi(8) = 4 \Rightarrow \varphi(4) = 2 \Rightarrow \varphi(2) = 1$ and $1 + 2 + 4 + 8 \neq 24$

The first few perfect totient numbers are 3, 9, 15, 27, 39, 81, 111, 183, 243, 255, 327, 363, 471, ... (Sequence A082897 in the OEIS)

Note that all powers of 3 are perfect totient numbers since

$\varphi(3^k) = 3^{k-1}(3-1) = 2 \times 3^{k-1}$

(by applying $\varphi(p^k) = p^{k-1}(p-1)$, where p is a prime number)

If we re-apply Euler's totient function to this result, we get

64 Euler's Totient function

$$\varphi(2 \times 3^{k-1}) = \varphi(2) \times \varphi(3^{k-1}) = 1 \times \varphi(3^{k-1}) = \varphi(3^{k-1})$$
$$= 3^{k-2}(3-1) = 2 \times 3^{k-2}$$

Surely, you have noticed the pattern.

In general,

$$\varphi(2 \times 3^i) = \varphi(3^i) = 2 \times 3^{i-1}.$$

So, repeating this procedure until reaching 1 yields many terms whose sum is $S = 1 + 2 \times 1 + 2 \times 3 + 2 \times 3^2 + 2 \times 3^3 + \ldots + 2 \times 3^{i-1}$

$$= 1 + 2(1 + 3 + 3^2 + 3^3 + \cdots + 3^{k-1})$$

Or, $1 + 3 + 3^2 + 3^3 + \cdots + 3^{k-1}$ is a sum of the first k terms of a geometric sequence (the sequence is 3^n and the sum is $\sum_0^{k-1} 3^k$). And it equals $1 \times \frac{3^k - 1}{3 - 1} = \frac{3^k - 1}{2}$.

Hence, $S = 1 + 2 \times \frac{3^k - 1}{2} = 1 + 3^k - 1 = 3^k$. Therefore, 3^k is a perfect totient number.

Carmichael's totient function conjecture

Carmichael's totient function conjecture is an important open problem that involves Euler's totient function.

In 1907, American mathematician **Robert Carmichael** stated that **for every integer n, there exist at least an integer m ≠ n such that the equation $\varphi(m) = \varphi(n)$ has k solutions.** [6]

For example, if n = 3, then the possible values of m are 4, 5, 6, … because $\varphi(3) = \varphi(4) = \varphi(5) = \varphi(6) = 2$.

You can check some sequences related to this function on OEIS' website (oeis.org)

- Numbers n such that φ(n) = k for different values of k: sequence A032447 in the OEIS
- The number of numbers m with Euler phi(m) = n: sequence A014197 in the OEIS

This conjecture can be stated differently as :

If $N_\varphi(k)$ is the number of positive integers n such that $\varphi(n) = k$, then $N_\varphi(k) \neq 1$.

At first, Carmichael announced this statement as a theorem. However, it turned out that his proof contained a flaw. So, it has been a conjecture since 1922. [7] To date, this conjecture is still an open problem [8]. Will you give it a shot?

In 1998, American mathematician **Kevin Ford** showed that the smallest counterexample to this conjecture, if it exists, must be greater than $10^{10^{10}}$ This number has more than 10^{10} digit. [9]

What is interesting about this function is that if one counterexample is found, then there must exist infinitely many of them. [10]

An interesting theorem, proved by Kevin Ford in 1998 and related to this conjecture is called "Sierpiński's conjecture" (even though it was proved) states that any positive integer greater than or equal to 2 is a number of solutions to the equation $\varphi(n) = k$, where n and k are positive integers. In other words, all the positive integers occur as values of the totient valence function. The totient valence function, usually denoted by N_φ, of an integer m, $N_\varphi(m)$, is the number of integers n for which $\varphi(n) = m$. [11+12]

A wonderful result concerning this last conjecture was proved by Hungarian mathematician **Paul Erdős** (1913 –1996) and it states that all values of N_φ, also called multiplicities, occur infinitely often. [13]

Lehmer's Totient Problem

Lehmer's Totient Problem is an interesting problem that involves the totient function.

In 1932, American mathematician **Derrick Henry Lehmer** (1905 – 1991) asked whether there exist composite numbers n such that $\varphi(n)$ divides n-1. Of course, if p is a prime number then $\varphi(p)$ divides p-1. However, no such composite numbers are known. [14+15+16]

In the same year, Lehmer showed that n must be odd, square-free, and divisible by at least 7 distinct prime numbers.

> A number is said to be **square-free** if its prime decomposition does not contain any repeated factors. In other words, it is not divisible by any perfect square other than 1.

It was also proved that n has to be a **Carmichael Number** (learn more about those numbers in the chapter covering Fermat's work).

It is known that if such numbers exist, then the smallest will be greater than 10^{22} and it has to be divisible by at least 14 distinct prime numbers. [15]

Chapter - 9

Mersenne Primes

Mersenne primes are prime numbers of the form $2^p - 1$, where p is a prime number.

They are generally denoted by M_p.

In the general case, a number of the form $2^n - 1$, with n being a positive integer, is called a Mersenne number.

Note:

> If $2^n - 1$ is prime then so is n. However, n being prime is a necessary, but not a sufficient condition for $2^n - 1$ to be prime. The smallest counterexample is for p = 11, $M_{11} = 2^{11} - 1 = 2047 = 23 \times 89$.

The fact stated above is easy to prove using a simple factorization trick. If n is not prime, then it has to be composite. Thus, we can write n as a product of two non-zero integers a and b. Therefore,

$2^n - 1 = 2^{a \times b} - 1$

$\qquad = (2^a)^b - 1$

$\qquad = (2^a - 1)[\,(2^a)^{b-1} + (2^a)^{b-2} + \cdots + (2^a)^1 + (2^a)^0\,]$

$\qquad = (2^a - 1)[\,(2^a)^{b-1} + (2^a)^{b-2} + \cdots + (2^a)^1 + 1\,]$

68 Mersenne Primes

Similarly,

$$2^n - 1 = 2^{a \times b} - 1$$
$$= (2^b)^a - 1$$
$$= (2^b - 1)[\,(2^b)^{a-1} + (2^b)^{a-2} + \cdots + (2^b)^1 + (2^b)^0\,]$$
$$= (2^b - 1)[\,(2^b)^{a-1} + (2^b)^{a-2} + \cdots + (2^b)^1 + 1\,]$$

Since the 16th century, many mathematicians have started to investigate the primeness of the numbers $2^n - 1$. They believed that those numbers are primes for any positive integer n. However, in 1536, **Hudalricus Regius**, showed that $2^{11} - 1$ (the counterexample included earlier) is composite. [1]

Those numbers are called after **Marin Mersenne,** a French mathematician who published in 1644 a book titled *"Cogitata Physica-Mathematica "* in which he stated that the numbers $2^n - 1$ are primes for n = 2, 3, 5, 7, 13, 17, 19, 31, 67, 127, 257 and composite for all the other integers less than 257. However, the biggest Mersenne prime that was known until then is M_{19}, was verified in 1603 by Italian mathematician **Pietro Cataldi [2+3]**. So, it was a conjecture that M_{31}, M_{67}, M_{127} and M_{257} are primes. In fact, Mersenne thought that $2^p - 1$ is prime if and only if p is a prime number that can be written in one of the following forms: $2^{2k} + 1$, $2^{2k} \pm 3$ or $2^{2k+1} - 1$, with k a positive integer. He believed that this was a necessary and a sufficient condition. [4]

Figure 1:
Marin Mersenne
(1588-1648)

However, this assumption was flawed. A simple counterexample is when p = 3. In this case, we have $2^3 - 1 = 7$ which is prime. Moreover, M_p is composite for p = 257 = $2^8 + 1$ and p = 67 = $2^6 + 3$. Mersenne's list was also incomplete because M_{61}, M_{89} and

M_{107} are prime numbers. It wasn't until 1947 that all the range below 257 was verified. Thus, M_n is composite for all positive integers less than 257 except for n being 2, 3, 5, 7, 13, 17, 19, 31, 61, 89, 107, or 127. [5]

An interesting theorem related to Mersenne primes and that a curious mind may be pleased to know is the following:

If $2^n - 1$ is a Mersenne prime, then n must be a prime number.

If you are wondering about the proof, here it is:

The theorem mentioned above states that if $2^n - 1$ is prime, then so is n. So, by way of contradiction, let's suppose that $2^n - 1$ is prime and n is composite. Thus, n can be written as the product of two positive non-zero integers a and b. So, n = a × b.

Hence, we have:

$2^n - 1 = 2^{ab} - 1 = (2^a)^b - 1$

$= (2^b - 1)[\,(2^b)^{a-1} + (2^b)^{a-2} + \cdots + (2^b)^1 + (2^b)^0\,]$

$= (2^b - 1)[\,(2^b)^{a-1} + (2^b)^{a-2} + \cdots + (2^b)^1 + 1\,]$

Notice that $2^n - 1$ is composite since it is the product of two integers which is a contradiction because $2^n - 1$ is prime. Therefore, n must be prime.

In this case, the first factor of n has to be 1 since the second one is strictly greater than 1.

The connection between Sophie Germain primes and Mersenne primes

In the 19th, one of the first female mathematicians emerged as an important figure that helped shape the foundation of two different areas of mathematics: number theory and mathematical physics. French mathematician Sophie Germain never attended school and was excluded from the learned societies. However, her scientific contributions and the methods she created are genuinely incredible.

Mersenne Primes

Her name will be remembered forever, especially when talking about Safe and Sophie Germain primes, a certain kind of prime numbers called after her and that she used in her investigations into the famous Fermat's Last Theorem.

A Sophie Germain prime is a prime number p such that 2p+1 is also prime. 2p+1 is called a safe prime, in other words, the prime number associated with a Sophie Germain prime is called a safe prime.

Figure 2:
Sophie Germain
(1776 - 1831)

For example, 5 is a Sophie Germain prime since $5 \times 2+1 = 11$ is also prime.

Around 1825, Sophie Germain proved that the first case of Fermat's Last Theorem is true for such primes. Soon after, Legendre began to generalize this by showing the first case of FLT also holds for odd primes *p* such that *kp*+1 is prime, *k*=4, 8, 10, 14, and 16. (If you are not familiar with Fermat's Last Theorem, you can check it in the next chapter.)

[6+7]

The first few Sophie Germain primes and safe primes are:

Sophie Germain primes:

2, 3, 5, 11, 23, 29, 41, 53, 83, 89, 113, 131, 173, 179, 191, 233, 239, 251, 281, 293, 359, 419, 431, 443, 491, 509, 593, 641, 653, 659, 683, 719, 743, 761, 809, 911, 953, ... (OEIS: A005384)

Safe primes

5, 7, 11, 23, 47, 59, 83, 107, 167, 179, 227, 263, 347, 359, 383, 467, 479, 503, 563, 587, 719, 839, 863, 887, 983, 1019, 1187, 1283, 1307, 1319, 1367, 1439, 1487, 1523, 1619, 1823, 1907, ... (OEIS: A005385)

As of June 2020, the biggest Sophie Germain prime is
$$2618163402417 \cdot 2^{1290000} - 1.$$

Discovered in February 2016, this number is 388,342 digits long. [8]

After reading about Sophie Germain primes, you may be thinking that if p is a Sophie Germain Prime, then 2p+1 is also prime, but what if 2p+1 is also a Sophie Germain Prime, then 2(2p+1)+1 is prime, and so on... By repeating this procedure, you will get a sequence or a chain of Sophie Germain primes that are related to each other. Such a chain is called a **Cunningham chain.** For example, {2, 5, 11, 23, 47} is a Cunningham chain of length 5.

In fact, there are two kinds of Cunningham chains: a sequence of k primes, each which is twice the preceding one plus one, is called a **Cunningham chain of length k of the first kind**, however, if each prime in the sequence is twice the preceding one minus one, then it is a **Cunningham chain of length k of the second kind**. For example, {2, 3, 5} and {1531, 3061, 6121, 12241, 24481} are Cunningham chains of lengths 3 and 5, respectively, of the second kind. [9]

Interestingly, Sophie Germain primes are connected to Mersenne primes.

In fact, there is a theorem in number theory that reveals this connection. Announced at first by Euler in 1750 and proved 25 years later by Lagrange, the theorem states that:

If p is a prime number different from 3 and of the form 3+4k, with k an integer, in other words, p ≡ 3 (mod 4), then 2p+1 is prime if and only if 2p+1 divides M_p.

(In this case, p is called a Sophie Germain prime).

The proof of this theorem may require the knowledge of some concepts in modular arithmetic such as the **Quadratic Residue** and the **Multiplicative order**. However, the proof won't be included in this book. Feel free to search for it on your own. (see [10] in the References section for the proof).

72   Mersenne Primes

If you are not familiar with the **Quadratic Residue** and the **Multiplicative order**, here is a small explanation of what they are.

Quadratic residue: If a is congruent to the square of b modulo c, with a, b, and c are integers, then a is called a quadratic residue modulo c. If not, a is called a quadratic non-residue modulo c.

For example, $5^2 \equiv 2 \ (mod\ 23)$. Therefore, 2 is called a quadratic residue modulo 23

Multiplicative order: The order of a modulo b is the smallest non-zero positive integer such that $a^n \equiv 1 \ (mod\ b\ )$.

For example, let a = 3 and b = 7.
In order to determine the order of 3 modulo 7, we can just substitute n for 1, then 2, then 3, … until we get to 1. So,

$3^1 \equiv 3 \ (mod\ 7\ )$

$3^2 \equiv 2 \ (mod\ 7\ )$

$3^3 \equiv 6 \ (mod\ 7\ )$ Therefore, the order of 3 modulo 7 is 6.

$3^4 \equiv 4 \ (mod\ 7\ )$

$3^5 \equiv 5 \ (mod\ 7\ )$

$3^6 \equiv 1 \ (mod\ 7\ )$

A quick trick:

When calculating exponentials in modular arithmetic, you can use the property that if $a \equiv b \ (mod\ c\ )$ and $a' \equiv b' \ (mod\ c\ )$, then $a \times a' \equiv b \times b' \ (mod\ c\ )$. For example, instead of calculating $3^6 = 729$ and then dividing by 7 and taking the remainder, you can just use the fact that (assuming you calculated it before) $3^5 \equiv 5 \ (mod\ 7\ )$.

So, $3^6 \equiv 3 \times 3^5 \equiv 3 \times 5 \equiv 15 \equiv 1 (mod\ 7\ )$. Using this trick will make the calculations easier, especially when the numbers get larger and larger.

Hunting for Mersenne primes

For hundreds of years, finding the next Mersenne prime was a laborious and time-consuming task. Verifying the primeness of M_{89} is a matter of some milliseconds using a computer, however, this wasn't that easy in the past.

Since the invention of computers, we have become able to do some onerous calculations faster than anyone could imagine. Calculations such as checking the primeness of new numbers and reverifying the ones we know already have become a lot easier and more popular due to the continuous advancements in the computer hardware industry. From the 60s to the 90s, scientists and engineers had the privilege of writing and executing the codes, used in those calculations, on the most powerful computers available then, and that were installed generally in laboratories and research centers. However, in 1995, a computer scientist called **George Woltman** founded the **Great Internet Mersenne Prime Search (GIMPS)** which is a distributed computing project consisting of volunteers that run, on their personal computers, a small program designed by GIMPS to search for the next biggest Mersenne prime and to reverify existing ones. This project was revolutionary since it gave the chance to enthusiasts to hunt for the next biggest Mersenne prime using highly optimized algorithms.[11]

In fact, the last 17 Mersenne primes known to date (August 2020) were discovered by volunteers participating in this project.

The last one discovered (and it is the biggest prime known as of August 2020) is the 24,862,048-digit long $M_{82,589,933}$. To date, only 51 Mersenne primes are known.

If you are interested in this project, you can read more about it on its website: www.mersenne.org

The following is a list of all the known Mersenne primes as of August 2020:

Mersenne Primes

#	2^p-1	Digits	Date Discovered	Discovered By
1	2^2-1	1	c. 500 BCE	Ancient Greek mathematicians
2	2^3-1	1	c. 500 BCE	Ancient Greek mathematicians
3	2^5-1	2	c. 275 BCE	Ancient Greek mathematicians
4	2^7-1	3	c. 275 BCE	Ancient Greek mathematicians
5	$2^{13}-1$	4	1456	Anonymous
6	$2^{17}-1$	6	1588	Pietro Cataldi
7	$2^{19}-1$	6	1588	Pietro Cataldi
8	$2^{31}-1$	10	1772	Leonhard Euler
9	$2^{61}-1$	19	1883	Ivan Mikheevich Pervushin
10	$2^{89}-1$	27	1911 Jun	R. E. Powers
11	$2^{107}-1$	33	1914 Jun 11	R. E. Powers
12	$2^{127}-1$	39	1876 Jan 10	Édouard Lucas
13	$2^{521}-1$	157	1952 Jan 30	Raphael M. Robinson
14	$2^{607}-1$	183	1952 Jan 30	Raphael M. Robinson
15	$2^{1,279}-1$	386	1952 Jun 25	Raphael M. Robinson
16	$2^{2,203}-1$	664	1952 Oct 07	Raphael M. Robinson
17	$2^{2,281}-1$	687	1952 Oct 09	Raphael M. Robinson
18	$2^{3,217}-1$	969	1957 Sep 08	Hans Riesel
19	$2^{4,253}-1$	1,281	1961 Nov 03	Alexander Hurwitz
20	$2^{4,423}-1$	1,332	1961 Nov 03	Alexander Hurwitz
21	$2^{9,689}-1$	2,917	1963 May 11	Donald B. Gillies
22	$2^{9,941}-1$	2,993	1963 May 16	Donald B. Gillies
23	$2^{11,213}-1$	3,376	1963 Jun 02	Donald B. Gillies
24	$2^{19,937}-1$	6,002	1971 Mar 04	Bryant Tuckerman
25	$2^{21,701}-1$	6,533	1978 Oct 30	Landon Curt Noll & Laura Nickel
26	$2^{23,209}-1$	6,987	1979 Feb 09	Landon Curt Noll
27	$2^{44,497}-1$	13,395	1979 Apr 08	Harry Lewis Nelson & David Slowinski
28	$2^{86,243}-1$	25,962	1982 Sep 25	David Slowinski
29	$2^{110,503}-1$	33,265	1988 Jan 28	Walter Colquitt & Luke Welsh
30	$2^{132,049}-1$	39,751	1983 Sep 19	David Slowinski

Mersenne Primes

#	Prime	Digits	Date	Discoverer
31	$2^{216,091}-1$	65,050	1985 Sep 01	David Slowinski
32	$2^{756,839}-1$	227,832	1992 Feb 19	David Slowinski & Paul Gage
33	$2^{859,433}-1$	258,716	1994 Jan 04	David Slowinski & Paul Gage
34	$2^{1,257,787}-1$	378,632	1996 Sep 03	David Slowinski & Paul Gage
35	$2^{1,398,269}-1$	420,921	1996 Nov 13	GIMPS / Joel Armengaud
36	$2^{2,976,221}-1$	895,932	1997 Aug 24	GIMPS / Gordon Spence
37	$2^{3,021,377}-1$	909,526	1998 Jan 27	GIMPS / Roland Clarkson
38	$2^{6,972,593}-1$	2,098,960	1999 Jun 01	GIMPS / Nayan Hajratwala
39	$2^{13,466,917}-1$	4,053,946	2001 Nov 14	GIMPS / Michael Cameron
40	$2^{20,996,011}-1$	6,320,430	2003 Nov 17	GIMPS / Michael Shafer
41	$2^{24,036,583}-1$	7,235,733	2004 May 15	GIMPS / Josh Findley
42	$2^{25,964,951}-1$	7,816,230	2005 Feb 18	GIMPS / Martin Nowak
43	$2^{30,402,457}-1$	9,152,052	2005 Dec 15	GIMPS / Curtis Cooper & Steven Boone
44	$2^{32,582,657}-1$	9,808,358	2006 Sep 04	GIMPS / Curtis Cooper & Steven Boone
45	$2^{37,156,667}-1$	11,185,272	2008 Sep 06	GIMPS / Hans-Michael Elvenich
46	$2^{42,643,801}-1$	12,837,064	2009 Jun 04	GIMPS / Odd M. Strindmo
47	$2^{43,112,609}-1$	12,978,189	2008 Aug 23	GIMPS / Edson Smith
48*	$2^{57,885,161}-1$	17,425,170	2013 Jan 25	GIMPS / Curtis Cooper
49*	$2^{74,207,281}-1$	22,338,618	2016 Jan 07	GIMPS / Curtis Cooper
50*	$2^{77,232,917}-1$	23,249,425	2017 Dec 26	GIMPS / Jon Pace
51*	$2^{82,589,933}-1$	24,862,048	2018 Dec 07	GIMPS / Patrick Laroche

* Provisional ranking, not all candidates between M43,112,609 and M82,589,933 have been eliminated.

[12]

How to find Mersenne primes?

This is an interesting question. Well, you can guess and check.

If you are still curious to know if there exists a quicker way, I am pleased to tell you that the answer is yes, there exists another method and it is based on a theorem in number theory called the **Lucas-Lehmer Theorem**.

It states that:

For any odd prime number p, $M_p = 2^p - 1$ is prime if and only if M_p divides S(p-1), where S(n) is a sequence satisfying the following conditions $S(n+1) = S(n)^2 - 2$ and $S(1) = 4$.

However, the proof of this theorem is beyond the scope of this book.

It was the French mathematician **François Édouard Anatole Lucas** who worked on this problem in 1870, and his work was later improved by American mathematician **Derrick Henry Lehmer**.

The theorem stated previously is indeed important since it is the essence of the **Lucas-Lehmer Primality test.**

Figure 3:
Derrick Henry Lehmer
(1905-1991)

Figure 4:
Édouard Lucas
(1842-1891)

If you are interested in applying what you have just learned, you can try to implement it into a computer program.

Interesting facts about Mersenne primes

- Is there an infinite number of Mersenne primes?

    The answer to this question is still unknown. Some conjectures that claim that this is true, but they have not yet been proved.

- This fact may not be related directly to Mersenne primes, but did you know that the **Electronic Frontier Foundation** which is an international non-profit digital rights organization, has offered four cash prizes to "encourage innovative computing that brings together ordinary Internet users to collectively contribute to solving huge scientific problems". [13]
    This organization offered $50,000 for the first individual or group that discover(s) the first 1,000,000 digit-long prime number. On April 6,2000, this prize was given to **Nayan Hajratwala** of Plymouth, Michigan, a participant of the GIMPS project. The next prize was worth $100,000 and it went to GIMPS project for the collaborative discovery of the first 12 million digit-long prime number, exceeding the minimum length required for the prize by 2 million digits. However, you still have a chance, dear reader, to win $150,000 or $250,000 for discovering the first 100,000,000 and 1 billion decimal digit-long primes, respectively. [13]

- Have you ever used the random function in Microsoft Excel? Or maybe you used it when coding in your favorite programming language like Python, PHP, R, Ruby, ... [14]

    If you are not familiar with the random function, in simple words, it is a built-in function in many software systems that gives you a "random" value. Technically speaking and without diving deep into the details, the value generated is not indeed random, but rather a pseudo-random. For the computer to do so, it uses a Pseudo-Random Number Generator (PRNG), which an algorithm that is based on mathematical formulas. One of the most famous PRNGs is the **Mersenne Twister**. It is called so because it uses Mersenne primes and especially $M_{19,937}$ as its period length. This means that it generates almost $M_{19,937}$ value before repeating itself. [15]

- Are you interested in recreational mathematics? If your answer is yes, I am sure you know, or at least heard about the famous **Tower of Hanoi**. If not, this is just a classic game of logical thinking and sequential reasoning that was invented in 1883 by French mathematician **Édouard Lucas.**
The toy set consists of three rods fastened to a stand and of a number of disks (often it is 8). The disks, each with a different radius, are initially placed on one of the pegs, such that the largest disk is on the bottom and the smallest on top. The player is asked to transfer the stack from the current rod to one of the others following two rules: only one disk can be moved at a time, and no disk can be placed on a smaller one. [16]

Figure 5: Tower of Hanoi

The minimum number of steps required to move n disks from one peg to another on is $2^n - 1$, a Mersenne number. I suggest that you, dear reader, try to prove this result. [17]

Perfect numbers

Perfect numbers have been known since antiquity because many civilizations believed in their mystic powers.

A number is said to be perfect if it equals the sum of its divisors excluding itself.

Mersenne Primes

For example, 6 is a perfect number (and it is, by the way, the smallest) because the sum of its divisors excluding 6, which are 1, 2, and 3, equals 6.

The next perfect number is 28 because $1+2+4+7+14 = 28$.

The next two are 496 and 8128.

Interestingly, perfect numbers have a strong connection with Mersenne primes. In fact, a perfect number can be written in the form of

$2^{n-1}(2^n - 1)$ with $2^n - 1$ being a Mersenne prime.

This is indeed a theorem in number theory that is called Euclid-Euler's theorem and it states that:

> **n is an even perfect number if and only if it has the form of $2^{k-1}(2^k - 1)$ with $2^k - 1$ is a Mersenne prime.**

It was Euclid the first to prove that if $2^n - 1$ is a prime number then $2^{n-1}(2^n - 1)$ is a perfect number. Almost 2000 years later, Euler proved the converse of this theorem. [18]

Notice that this theorem deals only with even perfect numbers. To date, it is still unknown if any odd perfect number exists. If there exists any, we are sure that the smallest one has more than 300 decimal digits[19] and is divisible by a prime number bigger than 10^{20}. [20]

Proof

Before getting to the proof, we will learn about a new function that may make our work easier.

This function is called the **sigma function** and it is denoted σ(n).

σ(n) is equal the sum of all the divisors of n, with n a non-zero positive integer.

For example, $\sigma(1) = 1$, $\sigma(2) = 1+2 = 3$, $\sigma(10) = 1+2+5+10 = 18$, ...

Obviously, if p is a prime number, $\sigma(p) = p+1$.

Mersenne Primes

Notice that one of the properties of this function is that it is multiplicative.

A function **f** is said to be multiplicative, if $f(m \times n) = f(m) \times f(n)$ for any integers m and n.

Let's go back to our proof.

By definition, a perfect number n is equal to the sum of its divisors excluding itself. So, we need to prove that for n to be a perfect number, $\sigma(n)$ must equal 2n (since it is the sum of all divisors of n including n itself).

We know that if $2^k - 1$ (with k a positive integer) is a prime number, therefore:

$\sigma(2^k - 1) = 2^k - 1 + 1 = 2^k$.

Or, $\sigma(2^{k-1})$ is the sum of all the divisors of 2^{k-1}. Hence,

$\sigma(2^{k-1}) = 2^0 + 2^1 + 2^2 + 2^3 + ... + 2^{k-3} + 2^{k-2} + 2^{k-1}$.

This is the sum of a simple geometric sequence. Using the summation notation, we can write it as follows:

$$\sum_{i=0}^{k-1} 2^i = 2^0 \times \frac{2^{k-1-0+1} - 1}{2 - 1} = 2^k - 1$$

Thus, based on the fact that σ is a multiplicative function, we get

$\sigma(n) = \sigma(2^{k-1}) \times \sigma(2^k - 1) = (2^k - 1) \times 2^k$

$= 2 \times [(2^k - 1) \times 2^{k-1}] = 2n$

Therefore, we can conclude that $2^{k-1}(2^k - 1)$ is a perfect number.

On the other hand, if n is an even perfect number, then we can write n as $2^{k-1}m$, with k is an integer greater than one and m is a positive odd integer. (We extracted all the powers of 2 from n).

So, σ(n) = σ(2^{k-1} × m) = σ(2^{k-1}) × σ(m) since σ is a multiplicative function

$$\sigma(n) = (2^k - 1) \times \sigma(m)$$

Or, n is a perfect number. Thus, σ(n) = 2n (as explained earlier).

Combining the last two results, we get

$$(2^k - 1) \times \sigma(m) = 2n = 2 \times 2^{k-1}m = 2^k m$$

Therefore, $(2^k - 1) \times \sigma(m) = 2^k m$ (A)

Using the generalized form of Euclid's Theorem that states:

If n divides a×b and n and a are relatively prime, then n must divide b.
[21]

And given that $2^k - 1$ divides $2^k m$, and the gcd $(2^k - 1, 2^k) = 1$ (**the gcd of two consecutive numbers is always equal to 1**), we can conclude that $2^k - 1$ must divide m. Thus, we can write m as the product of $2^k - 1$ and a positive non-zero integer q: m = $(2^k - 1)$ ×q
Substituting the value of m in (A) for $(2^k - 1)$ ×q and dividing both sides by $2^k - 1$, we get σ(m) = 2^kq.

Or, we know that σ(m) = m + q + B, with B the some of the other divisors of m.

$$m + q = m + \frac{m}{2^k-1} = m \times (1 + \frac{1}{2^k-1}) = m \times (\frac{2^k-1+1}{2^k-1}) = m \times (\frac{2^k}{2^k-1})$$

Substituting m for $(2^k - 1)$ ×q, we get: m + q = 2^kq.

Or, we proved earlier that $2^k q = \sigma(m)$. Therefore, we get

$$\sigma(m) = m + q = m + q + B.$$

Thus, B = 0. Hence, we get σ(m) = m + q which is impossible unless m is a prime number. In this case, we have q = 1.

82 Mersenne Primes

Or, we have: m = $(2^k - 1) \times q = 2^k - 1$. And voilà ! m is a Mersenne prime.

To conclude, if n is a perfect number, then it is of the form

$(2^k - 1) \times 2^k$, with k a positive integer, and $2^k - 1$ is a Mersenne prime.

∎

Always curious about learning more, you will certainly find the next fact fascinating.

Did you know that if you sum the digits of a perfect number, then sum the digits of the resulting number ... and you repeat this process until the result is a one-digit number, then you will get the same result each time (except when the initial number is 6). [4+22]

So, what is this number?

Well, let's find out. Take for example 28. The sum of its digits is 2+8 = 10. The sum of the digits of 10 is 1+0 = 1. Apparently, it is 1.

Let's check 496. Its sum of digits is 4+9+6 = 19. The sum of the digits of 19 is 1+9 = 10. The sum of the digits of 10 is 1+0 = 1.

Each time we repeat this procedure, which is called the iterated sum of digits of a perfect number, we end up with 1.

The first who made this observation was the Italian mathematician **Niccolò Fontana Tartaglia**. [23]

But why is this always true ?

Curious about proofs, you may have started to think why is that true.

Figure 6:
Niccolò Fontana Tartaglia
(1499/1500 - 1557)

Here is a simple way of proving this.

For the sake of clarity, let's define a new function, let it be s, such that s(n) is the sum of the digits of the positive non-zero integer n.
For example, s(2357) = 2+3+5+7 = 17.

It is easy to show that $s(n) \equiv n \pmod 9$ and here is why:

Without loss of generality, let n = abcd, where a, b, c, and d are the digits of n. (we will work just with a 4-digit number but the following is true for any integer n)

Thus, n = 1000×a + 100×b + 10×c + d

$\quad$ = 999×a + a + 99×b + b + 9×c + c +d

$\quad$ = 9(111×a + 11×b +c) + a + b + c + d

$\quad$ = 9(111×a + 11×b +c) + s(n)

Hence, we have:

$$s(n) = n - 9(111 \times a + 11 \times b + c).$$

Using one of the properties of modular arithmetic, we can write:

$$s(n) \equiv n - 9(111 \times a + 11 \times b + c) \equiv n \pmod 9.$$

Or, we know that n is a perfect number. Based on Euclid-Euler's theorem (that we stated and proved earlier),

$n = 2^{p-1}(2^p - 1)$, with p an integer greater than one, and $2^p - 1$ is a Mersenne prime. As a result of a theorem that we proved earlier and which states that if $2^n - 1$ is prime than so is n, then we can deduce that p is prime.

Modulo 6, p is congruent to 1 or 5 if p is greater than or equal to 5 and it may be congruent to 2 or 3 if it is equal to 2 or 3, respectively. However, it is obvious that p cannot be congruent to 4 modulo 6; otherwise, p will be divisible by 2.

Now, we will check case by case the value of n modulo 9 depending on p.

The first cas is p = 2. Thus,

$$n = 2^{2-1}(2^2 - 1) = 2 \times 3 \equiv 6 \pmod 9.$$

However, 6, as we stated earlier, is an exception.

The second case is p = 3. Hence, we have

$$n = 2^{3-1}(2^3 - 1) = 4 \times 7 = 28 \equiv 1 \pmod 9.$$

Thus, our theorem holds for p = 3, n = 28.

The third case is when $p \equiv 1 \pmod 6$. For the sake of clarity, let's agree on writing p in the following way: p = 1 + 6k, where k is a positive non-zero integer.

Or, the order of 2 modulo 9 is 6 since it is the smallest integer such that $2^6 \equiv 1 \pmod 9$.

So,

$$n \equiv 2^{p-1}(2^p - 1) \equiv 2^{6k}(2^{6k+1} - 1) \equiv 2^{12k+1} - 2^6 \pmod 9$$

$$\equiv 2 \times (2^6)^{2k} - (2^6)^k \equiv 2 \times 1 - 1 \equiv 1 \pmod 9.$$

Repeating the same steps for $p \equiv 5 \pmod 6$, we will get $n \equiv 1 \pmod 9$.

With that, the proof is finished. ∎

Chapter 10

Pierre de Fermat

Pierre de Fermat

(1607–1665)

> *" It is impossible for any number which is a power greater than the second to be written as a sum of two like powers. I have a truly marvelous demonstration of this proposition which this margin is too narrow to contain. "*
>
> **Pierre de Fermat**

Pierre de Fermat is one of the most brilliant and productive mathematicians of his time. Independently or sometimes with other mathematicians, he laid the foundations of multiple areas of mathematics and contributed to others.

Pierre de Fermat was born on August 17, 1601, in Beaumont-de-Lomagne, France, and died on January 12, 1665, in Castres.

He was the son of a wealthy leather merchant and a consul of Beaumont-de-Lomagne. He received his primary education at a local Franciscan school. Then, he went to the University of Toulouse. Afterward, he moved to Bordeaux where he started working on mathematical problems and making many connections with famous mathematicians such as **Beaugrand** and especially **Etienne d'Espagnet**, who shared his mathematical interests. From Bordeaux, he moved to Orléans where he studied the law at the University of Orléans. He became later a lawyer and magistrate serving in the *Parlement of Toulouse*. [1]

Since his childhood, Fermat had a gift for languages. He was fluent in classical Greek, Latin, Italian, Spanish, and Occitan. This skill helped him in his scientific career since he was able to read the most important scientific books and to become acquainted with the latest mathematical discoveries of his time. [1]

Fermat was a prolific scientist. His contributions to mathematics included the development of the analytic geometry in 1629 in his *"Method for determining Maxima and Minima and Tangents for Curved Lines and Introduction to Plane and Solid Loci"*, predating the publication of Descartes' famous *La géométrie (1637)*. However, since his work was published posthumously in 1679, the credit for inventing the analytic geometry went to Descartes. Indeed, Fermat discovered that applying algebra to some geometric problems through a coordinate system made them easier to solve, a method called, however, Cartesian Geometry after Descartes.

From the analytic geometry, he went to study tangents to different kinds of curves, extrema (maxima and minima) of multiple functions, etc. Through this work, he laid the foundations of the infinitesimal calculus, a new branch of mathematics at that time. His ideas inspired later the renowned physicist **Isaac Newton** and the German mathematician **Gottfried Leibniz** to develop, independently, the differential calculus as we know it today.

Besides, Fermat was among the first mathematicians to study the modern theory of probability through some letters that he exchanged with **Blaise Pascal** in 1654 and in which they discussed the mathematics behind some games of chance. [1+3]

Fermat did not shine only in mathematics but also in physics, specifically in optics. He proved, using the methods of determining the extrema of functions, the **principle of least time**, sometimes called also **Fermat's principle**. Without getting into the details, this principle states that the light always travels between two points along the path of the shortest time. From this principle, Fermat could deduce the fundamental laws of reflection and refraction. [2+4]

Nevertheless, Pierre de Fermat is best remembered for his marvelous work in number theory. **Diophantus'** *Arithmetica* inspired him to study diophantine equations, which are polynomial equations having 2 or more unknowns such that only their integer solutions are studied. The following are some examples of diophantine equations: $ax + by = 1$, $x^n + y^n = z^n$,... (where all the variables are positive integers)

While Diophantus was satisfied with finding a single solution regardless of it being integer or not, Fermat sought all the possible integral solutions.

Fermat studied a certain kind of diophantine equations that are of the form $x^2 - ny^2 = 1$, where n a positive non-square integer and only positive integer solutions are sought for x and y. Two trivial solutions to this equation are x = 1 and y = 0 or x = -1 and y = 0. To be more accurate, Fermat rediscovered this kind of equations that were studied since the VIIth century by Indian mathematicians such as **Brahmagupta**.

However, this kind of equation is called mistakenly Pell's equations (sometimes, it is called also Pell-Fermat equations). [5]

Fermat also discovered a new way of factorization which is based on the representation of an odd integer n as the difference between two squares:

$n = a^2 - b^2$. Thus, n can be factored as follows:

$n = (a - b)(a + b)$. Writing the initial formula differently, we get $a^2 - n = b^2$. So, by trying different values of a and subtracting n, we can check each time if the result is a square, and if it is, we can factorize n as explained above. There is a better way of doing this where we use modular arithmetic. This method is useful especially when implemented on a computer. Indeed, this method is the base of many modern factorization algorithms.

> *Tout nombre impair non quarré est différent d'un quarré par un quarré, ou est la différence de deux quarrés, autant de fois qu'il est composé de deux nombres, et, si les quarrés sont premiers entre eux, les nombres compositeurs le sont aussi.*

> *Any non-square odd number is different from a square by a square, or is the difference of two squares, as many times as it is composed of two numbers, and, if the squares are co-prime, the composing numbers are, too.*

> ***Oeuvres de Fermat***(1643), p256-257[6]

Brilliant as usual, Fermat invented a new kind of mathematical proof based on contradiction. It is often called "Proof By Infinite Descent" or occasionally "Fermat's Method Of Descent".

This proof shows that a given statement cannot hold for any positive integer by supposing that it is true for some natural number a, and then showing that this implies that it has to be true for a smaller number b. Then, you show that it has to be true for a smaller integer c and so on... leading to an infinite descent. Since a decreasing sequence of positive integers is finite, i.e., you will reach ultimately a lower bound that is bigger than or equal to 0, which is a contradiction.

Indeed, Fermat used this method to prove the non-solvability of the diophantine equation $x^4 - y^4 = z^2$, where x, y, and z are integers.

During his life, Fermat challenged many mathematicians to solve the latter equation, but he did not publish the proof himself as he usually did. It was his son who discovered it among his father's papers and published it posthumously. [7]

Speaking of which, the problem stated above, and proved by Fermat, is called after him, Fermat's Right Triangle Theorem and it has many formulations that I encourage you, dear reader, to learn about them because they are truly fascinating since they bring together different branches of mathematics.

Here are two of them:

- One formulation reveals a connection with what is known as "elliptic curves" which are, basically, the (plane algebraic) curves of the form $y^2 = x^3 + ax + b$. It states that the only rational points on the (elliptic) curve $y^2 = x(x-1)(x+1)$ are the trivial points (0,0), (1,0), and (-1,0). [8]
- There do not exist two Pythagorean triangles* in which the two legs of one triangle are the leg and hypotenuse of the other triangle.

*(a Pythagorean triangle is a triangle whose sides are the non-zero positive integers a, b, and c such that $a^2 + b^2 = c^2$).

In 1638, Fermat announced another interesting theorem, that has been called, since then, **Fermat's Polygonal Number Theorem**. It states that every positive integer is a sum of at most 3 triangular numbers, 4 square numbers, ... , generally, n "n-gonal " number. [9]

Polygonal numbers:

A polygonal number, also known as an n-gonal number, is a number that can be represented as dots arranged in the shape of a regular n-gon.

All polygonal numbers start with one, and in each step, the number of dots used to form the current polygonal number increases to obtain the next one, generating a well-defined sequence of integers. [10]

The following illustration may help you understand this procedure for triangular numbers:

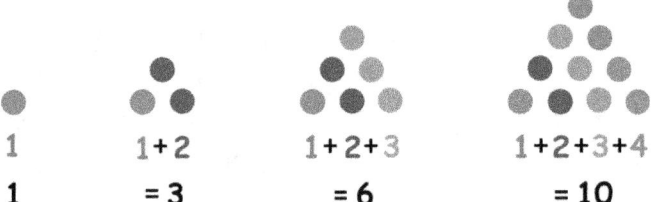

Figure 1: The first 4 triangular numbers

The first few triangular numbers are: 1, 3, 6, 10, 15,...
(Sequence A000217 in the OEIS)

There is indeed a formula that generates the nth triangular number, which is $\frac{1}{2}n(n+1)$. (Try to derive it on your own).

Another example of polygonal numbers is square numbers, which are numbers that can be arranged in a square shape.

This may be the easiest and the most intuitive one.

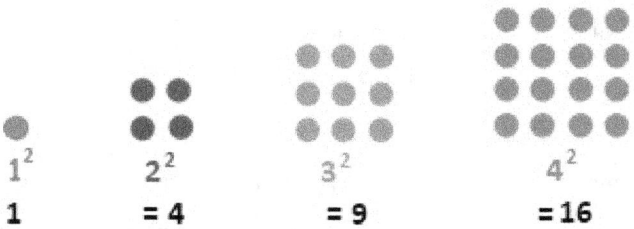

Figure 2: The first 4 square numbers

The formula that generates the nth square number is just n^2.

There is also the pentagonal numbers, hexagonal numbers, etc.

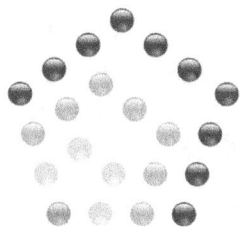

Figure 3: Pentagonal numbers

Polygonal numbers were discovered by the Greeks in c. 500 BC and this was due to their concrete way of thinking about numbers.

The Pythagoreans were the first to recognize that some numbers have geometric shapes.

However, some historical resources gave the credit of discovering the polygonal numbers to the Greek mathematician **Hypsicles** (c. 170-126 BC) known for his general formula of the nth a-gonal number. [11]

Speaking of which, here it is:

$$\frac{1}{2}n[2 + (n-2)(a-1)]$$

Here are some applications of this formula:

- Hexagonal numbers formula: $n(2n-1)$
- Heptagonal numbers formula: $\frac{1}{2}n(5n-3)$
- Enneadecagonal (19-gon) numbers formula: $\frac{1}{2}n(17n-15)$
- Megagonal (1,000,000-gon) numbers formula:
$$\frac{1}{2}n(999{,}998\,n - 999{,}996)$$

As he usually did, Fermat did not provide proof to his theorem of Polygonal Numbers.

In 1770, the Italian mathematician **Joseph Louis Lagrange** proved the second case of this problem which deals with square numbers. [12]

26 years later, German mathematician, **Carl Friedrich Gauss** proved the case of triangular numbers. He wrote in his diary:

" ΕΥΡΗΚΑ*! num = Δ + Δ + Δ " [12]

> * ΕΥΡΗΚΑ or Eureka is a Greek word that is used to celebrate a triumph or a discovery. Historically, this word is attributed to a popular legend about the Greek physicist and scholar **Archimedes** who proclaimed "Eureka! Eureka!" when he made the sudden realization that the buoyancy of an object placed in water is equal in magnitude to the weight of the water the object displaces.

Then, in 1813, French mathematician **Augustin-Louis Cauchy** proved the general case of the theorem. [12]

If you found this problem interesting, you may find Pollock's conjectures more fascinating because they are an extension of Fermat's theorem in the 3-dimensional space. Conjectured in 1850 by Sir **Frederick Pollock**, they state that every positive integer can be represented as the sum of at most 5 tetrahedral numbers (the first conjecture), and at most 7 octahedral numbers (second conjecture).

[13+14]

Tetrahedral numbers

A tetrahedral number is a number that can be represented using dots arranged in the form of a tetrahedron.

The first few tetrahedral numbers are:

1, 4, 10, 20, 35, 56, 84, 120, 165, 220, ... (A000292 in the OEIS)

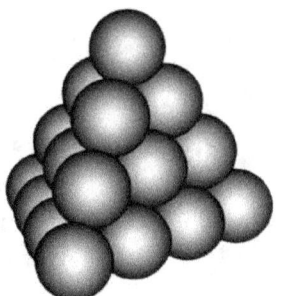

Figure 4: A tetrahedron with side length 4 contains 20 spheres. Each layer represents one of the first four triangular numbers.

Octahedral numbers

An octahedral number is a number that can be represented by dots arranged in the form of an octahedron.

The first few octahedron numbers are:

1, 6, 19, 44, 85, 146, 231, 344, 489, 670, 891, ...

(A005900 in the OEIS)

Figure 5: The 6th octahedral number: 146 magnetic balls, packed in the form of an octahedron with side length 6.

A generalized Pollock conjecture (the third conjecture), also known as the Polyhedral Numbers Conjecture, states that if m is the number of vertices of a platonic solid that is a "regular n-hedron" (n is 4, 6, 8, 12, or 20), then every positive integer is the sum of at most m+1 n-hedral numbers. To date, this conjecture still unproven.

So, let's generalize this problem more. Can a positive integer be represented as the sum of at most some pre-determined number of n-polytopial numbers (a regular n-polytope has (n-1) dimensional regular and congruent faces)? Think about it for a while. Overwhelming, isn't it?

Fermat's theorem on the sum of two squares

Among Fermat's contributions to the study of prime numbers, there is an interesting theorem that states that:

Any odd prime number p can be expressed as the sum of two squares, i.e., there exist two integers x and y such that $p = x^2 + y^2$, if and only if p is of the form 4k+1, where k is a positive non-zero integer.
(In other words, $p \equiv 1 \ (mod \ 4)$.)   [15+16+17]

Primes with this property are called Pythagorean Primes.

For example, the numbers 5, 13, 17, 29, 37, 41, 53, 61, ...
(sequence A002144 in the OEIS) are Pythagorean primes since

$5 = 1^2 + 2^2$, $13 = 2^2 + 3^2$, $17 = 1^2 + 4^2$, $29 = 2^2 + 5^2$, $37 = 1^2 + 6^2$, $41 = 4^2 + 5^2$, $53 = 2^2 + 7^2$, $61 = 5^2 + 6^2$, ...

At first, this was an observation made by French mathematician **Albert Girard** in 1625.   [18]

Girard studied all the positive integers that can be expressed as the sum of two squares. However, Fermat worked on an improved version of this observation that he sent in a letter to **Marin Mersenne**, dated 1640.
                [18+19]

Fermat's letter did not include any proof of his theorem. It wasn't until 1752 for the first proof to be published by Euler, where he used Fermat's Infinite Descent method. [20]

Fermat's Last Theorem

If you heard about Pierre de Fermat before, I am almost sure that it was for his Last Theorem. This theorem, as simple as it is, states that the diophantine equation $a^n + b^n = c^n$, where a, b, c, and n are positive integers, has no integer solutions for n > 2.

At first, this theorem was a conjecture and it remained so for a long time. It was called a theorem over the time because the great mathematician Fermat asserted it was correct and because there was no evidence of the contrary.

In opposition to what the name may refer to, this theorem wasn't the last Fermat stated, but instead, it was the last one that resisted a proof.

Obviously, the case where n = 1 has an infinite number of solutions. Similarly, if n = 2, all the Pythagorean triples* are solutions to this equation. In this case, there are also infinitely many solutions because the number of Pythagorean triples is infinite.

Pythagorean triples

Pythagorean triples are, in simple words, three positive integers a, b, and c such that $a^2 + b^2 = c^2$. They are often denoted by (a, b, c).

It may seem obvious to you that this idea is derived from the Pythagorean Theorem.

The most popular Pythagorean triple is (3,4,5). The next one is (8,6,10).

> Primitive Pythagorean triples are a special kind of Pythagorean triples. A Pythagorean triple is said to be primitive if a, b, and c are relatively prime.
>
> There is indeed a formula for generating Pythagorean triples that uses two positive integers m and n as parameters, and it states that the triple (a,b,c) is Pythagorean if
>
> $$a = m^2 - n^2, b = 2mn, \text{ and } a = m^2 + n^2.$$

For more than 350 years, Fermat's Last Theorem remained a tough challenge for many great minds who had been trying for a long time to solve it.

Even though they couldn't, their work was a great contribution to the development of mathematics and has opened new horizons to mathematical researches.

Uncountable unsuccessful proofs were suggested by many mathematicians. For that reason, it was considered by the Guinness Book Of World Records as the most difficult mathematical problem.

That centuries had passed without finding a proof had led many mathematicians to suspect that Fermat was mistaken in thinking he actually had a proof.

It wasn't until late 1994 that the first complete solution was found by the British mathematician **Andrew Wiles.**

The first formal proof was published in 1995, but it contained a flaw and it was corrected and republished (by him) in 1997.

Wiles' discovery is an important milestone in the history of mathematics.

Figure 6: Andrew Wiles

His proof is based on elliptic curves. More specifically, it used a conjecture stated in 1957 by **Yutaka Taniyama** and **Gorō Shimura** and called after them, the **Taniyama–Shimura conjecture** that, when it was first announced, had no connection with Fermat's Last Theorem. The connection was revealed in the 1980s (1985 and 1986) by German mathematician **Gerhard Frey**, French mathematician **Jean-Pierre Serre,** and American mathematician **Ken Ribet.**

Explaining this theory requires advanced knowledge in topology and number theory. However, in simple words, Wiles partially proved this conjecture which was sufficient for him to deduce that Fermat's Last Theorem is correct. [21] A few years later, in 2001, **Brian Conrad, Fred Diamond**, **Richard Taylor,** and **Christophe Breuil** fully proved the Taniyama–Shimura conjecture that has been known, since then, as The Modularity Theorem.

In 2016, Wiles was awarded the Abel Prize *"for his stunning proof of Fermat's Last Theorem by way of the modularity conjecture for semistable elliptic curves, opening a new era in number theory".* [22]

When Fermat first suggested his famous problem, it was written in the margin of his copy of Diophantus' *Arithmetica*. He included that he had a brilliant proof, but the margin was too small to contain it.

> *Cubum autem in duos cubos, aut quadratoquadratum in duos quadratoquadratos & generaliter nullam in infinitum ultra quadratum potestatem in duos eiusdem nominis fas est dividere cuius rei demonstrationem mirabilem sane detexi. Hanc marginis exiguitas non caperet.*
>
> *It is impossible to separate a cube into two cubes, or a fourth power into two fourth powers, or in general, any power higher than the second, into two like powers. I have discovered a truly marvelous proof of this, which this margin is too narrow to contain.*
>
> **Pierre de Fermat [23]**

Many mathematicians doubted the existence of this proof because only a few methods were available to Fermat. Moreover, the methods that were used in Wiles' proof were developed around 300 years after Fermat's death.

Nevertheless, Fermat did publish only the proof of a special case when the exponent n is 4 using his Infinite Descent method.

It is worth mentioning that after Fermat's proof, the theorem only had to be proved for odd prime numbers since if n >2, then it is divisible by 4 or by an odd prime or by both. In other words, if the exponent n is composite, then n can be factored into some integers p and q such that p is an odd prime. Therefore, by proving that FLT holds for p, then we can prove that it holds also for n because if $A^p + B^p = C^p$ doesn't have any natural solutions then the equation

$(a^q)^p + (b^q)^p = (c^q)^p$ (where A = a^q, B = b^q, and C = c^q)

that is, $a^n + b^n = c^n$ doesn't have either. [24]

Therefore, mathematicians who attempted to solve FLT concentrated on the first and second cases of it.

The first case is when a, b, c, and the exponent n are coprime, and the second is when n divides at least one of a, b, and c.

In 1832, French mathematician **Sophie Germain** made a breakthrough by proving that if the exponent p is an odd Sophie Germain prime, i.e., both p and 2p+1 are primes, then $a^p + b^p = c^p$ has no natural solutions.

Germain verified also all the odd prime numbers less than 270. [25]

Her work was significant, especially her Sophie Germain primes that are used, nowadays, in various fields such as cryptography, and primality testing, etc.

German mathematician **Ernest Kummer**'s work in proving special cases of FLT is also important since it led him to discover a new kind of numbers known as the Ideal Numbers. [26]

However, explaining what his work was about is beyond the scope of this book.

Generalization of Fermat's Last Theorem

Fermat-Catalan Conjecture

One of the generalized forms of Fermat's Last Theorem is a conjecture stating that the Diophantine equation $a^n + b^m = c^r$ has finitely many positive integer solutions a, b, c, m, n, and r with distinct triplets of value (a^n, b^m, c^r) such that a, b, and c are coprime and m, n, and r satisfy $\frac{1}{m} + \frac{1}{n} + \frac{1}{r} < 1$. [27]

This conjecture is known as Fermat-Catalan conjecture. As of August 2020, only 10 solutions of the equation stated above are known. They are:

$$1^p + 2^3 = 3^2 \ (p \geq 2)$$
$$2^5 + 7^2 = 3^4$$
$$13^2 + 7^3 = 2^9$$
$$2^7 + 17^3 = 71^2$$
$$3^5 + 11^4 = 122^2$$
$$33^8 + 1549034^2 = 15613^3$$
$$1414^3 + 2213459^2 = 65^7$$
$$9262^3 + 15312283^2 = 113^7$$
$$17^7 + 76271^3 = 21063928^2$$
$$43^8 + 96222^3 = 30042907^2$$

[28]

In fact, this conjecture is not only a generalization of Fermat's Last Theorem but also of Catalan's conjecture (which explains the name), suggested by **Eugène Charles Catalan** (1814 – 1894) in 1844 and proved in 2002 by Romanian mathematician **Preda V. Mihăilescu**. [29+30]
Simply, Catalan's Conjecture states that:

The only positive integer solutions to $x^a - y^b = 1$, where a and b are integers bigger than one and x and y are positive non-zero integers, are x = 3, a = 2, y = 2, and b = 3, that is, $3^2 - 2^3 = 1$. [31]

Beal's Conjecture

American Banker and amateur mathematician Andrew Beal suggested in 1993 when he was working on a generalized form of Fermat's Last Theorem, a conjecture that has been, since then, called after him, and it states that:

If A, B, C, x, y, and z are positive integers such as x, y, and z are bigger than 2, satisfying $A^x + B^y = C^z$, then A, B, and C have a common prime factor.

In other words, the equation $A^x + B^y = C^z$ has no positive integer solutions such that x, y, and z are at least 3, and A, B, and C are coprime.

[32+33]

It is worth mentioning that if one of the exponents x, y, and z is 2, then there are many counterexamples to this conjecture such as $7^3 + 13^2 = 2^9$ and $1^m + 2^3 = 3^2$, where m is a positive integer.

At first, Beal generously offered $5,000 for a proof or a counterexample to his conjecture to inspire young people to be more interested in science in general, and in mathematics in particular.

Since 1993, the value of the prize has increased. As of August 2020, the prize is worth one million US dollars and is held in trust by the **American Mathematical Society (AMS).** If you have proved this conjecture, you can visit their online website (www.ams.org) for more information about the prize and the way of submitting the proof. It is indeed a good resource to learn more about it.

Fermat's Little Theorem

Among Fermat's exceptional mathematical contributions, his Little Theorem stands out as one of the most important theorems in number theory.

On October 18, 1640, Fermat sent a letter to his friend, the French mathematician **Bernard Frénicle de Bessy** (1604? – 1674), in which he stated that:

> *Tout nombre premier mesure infailliblement une des puissances - 1 de quelque progression que ce soit, et l'exposant de la dite puissance est sous-multiple du nombre premier donné - 1; et, après qu'on a trouvé la première puissance qui satisfait à la question, toutes celles dont les exposants sont multiples de l'exposant de la première satisfont tout de même à la question ... Et cette proposition est généralement vraie en toutes progressions et en tous nombres premiers; de quoi je vous envoierois la démonstration, si je n'appréhendois d'être trop long.*
>
> *Every prime number divides necessarily one of the powers minus one of any progression. After one has found the first power that satisfies the question, all those whose exponents are multiples of the exponent of the first one satisfy similarly the question... And this proposition is generally true for all series and for all prime numbers; I would send you a demonstration of it, if I did not fear going on for too long.*

Pierre de Fermat, 1640[34+35]

This theorem simply states that:

If p is a prime number and a is a positive integer, then p must divide $a^p - a$. (this is generally written using modular arithmetic as $a^p \equiv a \ (mod\ p)$ [36]

There is another version of this theorem which is indeed a special case of it and it states that:

If a is a positive integer and p a prime number that doesn't divide a, then p must divide $a^{p-1} - 1$. (or, $a^{p-1} \equiv 1 \ (mod\ p)$) [36]

And this is easy to prove. We know that, by Fermat's original statement, p divides $a^p - a = a(a^{p-1} - 1)$. Thus, by Euclid's Lemma, if p doesn't divide a, then it has to divide $a^{p-1} - 1$.

It is worth mentioning that Fermat announced the general statement of the theorem. However, the other one is frequently used. [36]

Note that the converse of this theorem is not always true. For instance, if you find two positive integers n and m such that $n^{m-1} \equiv 1 \pmod{m}$, then you cannot deduce directly that m is prime. Prime numbers don't surrender easily!

Take for example 2^{341}. Modulo 341, this number is congruent to 2. However, 341 is composite and can be factored into 11 and 31.

Numbers p with this property, i.e., p divides $a^{p-1} - 1$ where a and p are integers and p is not prime, are called pseudoprimes, to be more accurate, a Fermat pseudoprime to base a.

Those numbers are indeed connected to another kind of numbers that are called **Carmichael numbers,** which are named after American mathematician **Robert Daniel Carmichael**.

A number p is called a Carmichael Number if it is a Fermat pseudoprime to base a for every a strictly less than p and relatively prime with it. [37]

Nevertheless, American mathematician Dr. **Derrick Lehmer** formulated a theorem, that is often called Lehmer's Theorem and, occasionally, " the converse of Fermat's Little Theorem". It states that:

If there exist two positive integers a and p such that $a^{p-1} \equiv 1 \pmod{p}$ and for all prime numbers i that divides $p - 1$, we have $a^{\frac{p-1}{i}}$ is not congruent to 1 modulo p, then p is prime. [38]

Fermat, as he usually did, did not provide a proof to his Little Theorem, even though he promised to do so, but, apparently, it was too long, as he always claimed.

The first to discover a proof was the German mathematician **Gottfried Wilhelm Leibniz,** and that was in 1683, but he did not publish it. [39]

The first published proof, however, was by Euler in 1736. [40]

Not only did Euler prove Fermat's Little Theorem, but he also generalized it as he usually did, and came out with what is known today as Euler's Theorem that we talked about earlier. (Seen Euler's totient function).

As you may remember, it states that if a and n are positive integers such that gcd(a, n) = 1, i.e., a and n are relatively prime, then

$$a^{\phi(n)} \equiv 1 \ (mod \ n),$$

where $\phi(n)$ is Euler's totient function.

As you may have noticed, Fermat's Little Theorem becomes a special case of Euler's Theorem because if n is prime, then $\phi(n) = n - 1$. In this case, we have: $a^{n-1} \equiv 1 \ (mod \ n)$. [41]

A proof of Fermat's Little Theorem

As a skeptical reader may be wondering why is this true, it is my pleasure to provide a proof. [42+43]

This proof is included in G. H. Hardy's *An Introduction to the Theory of Numbers*. However, it was discovered by British mathematician **James Ivory**, and it goes as follows.

The idea behind this proof is simple. Let a and p be positive integers such that a and p are co-prime and let's consider a first set of numbers that contains a, 2a, 3a, ..., (p-1)a . If we reduce each one of these modulo p, then we will get a second set that is composed of 1, 2, 3, ..., p-1.

This result follows from the fact that none of the numbers a, 2a, 3a, ..., (p-1)a is divisible by p, and also that they are distinct after being reduced modulo p. The latter statement can be proved by way of contradiction by assuming that there exist two different elements from the first set, i × a and j × a, that are congruent modulo p.

In this case, $ia \equiv ja \ (mod \ p)$ implies that $i \equiv j \ (mod \ p)$ (we can divide by a since a is non-zero and it is not divisible by p). Or, i and j are bigger than one and less than or equal to p-1. Therefore, they have to be equal. This result is absurd since we supposed i × a and j × a are different. Therefore, we conclude our hypothesis is false, which is to say that ia and ja must be distinct when reduced modulo p.

Since the p-1 element from the first set are distinct when reduced modulo p, then the only possible set of solutions is 1, 2, 3, ..., p-1.

If we multiply all the elements of the first set and reduce them modulo p, we get the following,

$$a \times 2a \times 3a \times ... \times (p-1)a \equiv 1 \times 2 \times 3 \times ... \times (p-1) \ (mod \ p)$$

which is equivalent to,

$$(p-1)! \times a^{p-1} \equiv (p-1)! \ (mod \ p)$$

By dividing both sides by (p-1)! (which is correct for the same reasons discussed earlier), we get

$$a^{p-1} \equiv 1 \ (mod \ p)$$

∎

Fermat Quotient

Fermat Quotient is strongly connected to his Little Theorem.

Fermat Quotient of an odd prime p with base a, denoted by $q_p(a)$ is defined as:

$$q_p(a) = \frac{a^{p-1} - 1}{p}$$

Note that it is an integer because, by Fermat's Little Theorem, p divides $a^{p-1} - 1$. [44+45]

Among the properties of Fermat Quotient that were proved in 1850 by German mathematician **Gotthold Eisenstein [46]**, we can mention the following:

- If p is a prime number and a and b are positive integers such that p divide neither a nor b, then $q_p(a \times b) \equiv q_p(a) + q_p(a) \ (mod \ p)$
- If p is a prime number, then

$$q_p(p-1) \equiv 1 \ (mod \ p) \text{ and } q_p(p+1) \equiv -1 \ (mod \ p).$$

In short,
$$q_p(p \pm 1) \equiv \mp 1 \ (mod \ p)$$

Wieferich primes

Wieferich primes are a special kind of prime numbers.

A prime number p is said to be a Wieferich prime if p^2 divides $2^{p-1} - 1$, i.e., $2^{p-1} \equiv 1 \ (mod \ p^2)$ [47]

Wiefeich primes were first introduced in 1909 by German mathematician **Arthur Josef Wieferich** when he was working on the first case of Fermat's Last Theorem. He showed that if Fermat's Last Theorem has any solutions for an odd prime exponent p, then p has to be a Wieferich prime. More technically, he proved that if there exist an odd prime number p and three positive integers x, y, and z such that $x^p + y^p + z^p = 0$ and p doesn't divide the product xyz, i.e., p is coprime with each one of x, y, and z, then p^2 divides $2^{p-1} - 1$, hence, p is a Wieferich number. [48]

The following year, Russian mathematician **Dmitry Semionovitch Mirimanoff** proved the same case for base 3. In other words, if Fermat's Last Theorem is false for an odd prime exponent p, then p^2 divides $3^{p-1} - 1$, i.e., $3^{p-1} \equiv 1 \ (mod \ p^2)$. However, to date, only two such primes are known. They are 11 and 1006003. (OEIS A014127 [50]) [47+49]

More generally, if a is a positive integer and p an odd prime such that a < p and p^2 divides $a^{p-1} - 1$, then p is called a Wieferich prime to base a.

The table below contains the Wieferich primes to the first few bases:

a	Primes p such that $a^{p-1} \equiv 1 \pmod{p^2}$	OEIS sequence
1	2, 3, 5, 7, 11, 13, 17, 19, 23, 29, ... (All primes)	A000040
2	1093, 3511, ...	A001220
3	11, 1006003, ...	A014127
4	1093, 3511, ...	
5	2, 20771, 40487, 53471161, 1645333507, 6692367337, 188748146801, ...	A123692
6	66161, 534851, 3152573, ...	A212583
7	5, 491531, ...	A123693
8	3, 1093, 3511, ...	
9	2, 11, 1006003, ...	
10	3, 487, 56598313, ...	A045616
11	71, ...	

[51]

Wieferich primes to base 2 are, as you may have noticed, rare. Indeed, only two of them are known which are 1093 and 3511. (A001220 in OEIS) However, if there exists any, the smallest will be bigger than 4.97×10^{17}

[50]

The existence of an infinite number of Wieferich primes is still a conjecture to be proved. [52]

Wieferich primes have a strong connection with Fermat Quotient. In fact, if $q_p(2) \equiv 0 \pmod{p}$, where p is an odd prime, then $2^{p-1} \equiv 1 \pmod{p^2}$. Therefore, p is a Wieferich prime. [53]

This is because, by Fermat's Little Theorem, a prime number p divides $2^{p-1} - 1$, and if $q_p(2) = \frac{2^{p-1}-1}{p}$ is also divisible by p, then $2^{p-1} - 1$ is divisible by p^2.

A curious mind eager to know more may be pleased to learn about the Double Wieferich Prime Pairs.

A pair of prime numbers (p, q) is said to be a Double Wieferich Prime Pair if and only if $q^{p-1} \equiv 1 \pmod{p^2}$ and $p^{q-1} \equiv 1 \pmod{q^2}$.

To date, only 7 pairs are known and they are (2, 1093), (3, 1006003), (5, 1645333507), (5, 188748146801), (83, 4871), (911, 318917), and (2903, 18787) (sequence A282293 in OEIS)   [54+55]

Wieferich numbers

An odd positive integer n is said to a Wieferich number if the congruence $2^{\varphi(n)} \equiv 1 \ (mod \ n^2)$ holds, where φ is Euler's totient function. (This is easily proved using Euler's Theorem)

Therefore, Weiferich primes become a special case of Wieferich numbers because if n is prime then φ(n) = n - 1.

The first few Wieferich numbers are 1, 1093, 3279, 3511, 7651, 10533, 14209, 17555, 22953, 31599, 42627, 45643, 52665, …
(sequence A077816 in the OEIS)   [56]

Wieferich numbers can be generalized to any base a such that

$$a^{\varphi(n)} \equiv 1 \ (mod \ n^2)$$

[57]

Fermat Numbers

Pierre de Fermat was the first to study the numbers of the form $2^{2^n} + 1$, where n is a positive integer (it can be 0). Those numbers, called after him, Fermat Numbers, are often denoted by F_n. Particularly, if F_n is prime, then it is called a Fermat Prime.

The first few Fermat Numbers are:

$F_0 = 3$,

$F_1 = 5$,

$F_2 = 17$,

$F_3 = 257$,

$F_4 = 65537,$

$F_5 = 4294967297,$

$F_6 = 18446744073709551617,$

$F_7 = 340282366920938463463374607431768211457,$

$F_8 =$
115792089237316195423570985008687907853269984665640394575
840007913129639937,

$F_9 =$
134078079299425970995740249982058461274793658205923933777236
1443721764030073546976801874298166903427690031858186486050853
75388281194656994643364900608409 7,

$F_{10} =$
179769313486231590772930519078902473361797697894230657273430
081157732675805500963132708477322407536021120113879871393357
658789768814416622492847430639474124377767893424865485276302
219601246094119453082952085005768838150682342462881473913110
540827237163350510684586298239947245938479716304835356329624
22413 7217. (A000215 in OEIS)[58]

As you may have noticed, the first 5 Fermat Numbers are prime (from F_0 to F_4). Based on this observation, Fermat conjectured, in many letters dated to 1640, that he sent to fellow mathematicians, that all numbers of the form $2^{2^n} + 1$, where n is a positive integer, are primes. [59]

However, it turned out that this conjecture is false. In fact, around a century later, in 1732, the great Swiss mathematician **Leonhard Euler** proved that $F_5 = 2^{2^5} + 1 = 2^{32} + 1$ is composite and showed that it is divisible by 641.

A curious reader may be asking how Euler did it. Well, as always, I am pleased to provide a proof.

Discovered in 1732, Euler's proof is based on another theorem that he proved previously and it states that:

Any prime factor of F_n, where n is an integer greater than 2, is of the form $k \times 2^{n+1} + 1$, where k is a positive non-zero integer. [60]

(Later, French mathematician **François Édouard Lucas** showed that k has to be even, so this becomes $k \times 2^{n+2} + 1$) [60]

Before getting to Euler's proof, it is necessary to know some interesting properties of Fermat Numbers.

In fact, the n+1^{rst} Fermat Number can be expressed in terms of the first n Fermat numbers by the following recurrence relation

$$F_n = F_0 \times F_1 \times F_2 \times F_3 \times \ldots \times F_{n-1} \times F_n + 2$$

(The proof of this formula is easy. In fact, if

$$F_0 \times F_1 \times F_2 \times F_3 \times \ldots \times F_{n-1} = F_n - 2$$

then,

$$F_0 \times F_1 \times F_2 \times F_3 \times \ldots \times F_{n-1} \times F_n$$
$$= F_n^2 - 2F_n$$
$$= (2^{2^n} + 1)^2 - 2(2^{2^n} + 1)$$
$$= 2^{2 \times 2^n} + 2 \times 2^{2^n} + 1 - 2 \times 2^{2^n} - 2$$
$$= 2 \times 2^{2^n} + 1 - 2$$
$$= F_{n+1} - 2$$

Based on this property, we can prove that any pair of Fermat Numbers are relatively prime and this is indeed a theorem in number theory, called **Goldbach's Theorem.** [61]

Here is the proof:

Let F_i and F_j be a pair of Fermat Numbers such that, without loss of generality, $0 \leq i \leq j$, and let c be the gcd of F_i and F_j. Notice that c has to be odd since both F_i and F_j are also odd.

By construction, c divides F_i and F_j.

Or, $F_j = F_0 \times F_1 \times F_2 \times ... \times F_i \times ... \times F_{j-1} + 2$.

Thus, $F_j - 2 = F_i \times Q$, where $Q = F_0 \times F_1 \times F_2 \times ... \times F_{j-1}$.

Hence, F_i divides $F_j - 2$, and we know that c divides F_i, so we can deduce that c divides $F_j - 2$.

Therefore, since c divides both F_j and $F_j - 2$, then c must divide $F_j - (F_j - 2) = 2$, and given that c is odd, then c has to be 1.

(The former statement is based on a fundamental property of the gcd of two integers that states that if d is the gcd of two integers a and b, then d divides $\alpha \times a + \beta \times b$, where α and β are two integers that can be positive, negative, or 0.)

As the gcd of F_i and F_j is one, then they are co-prime. [62]

As stated earlier, Euler proved that any prime factor of F_n must be of the form $k \times 2^{n+1} + 1$. So, all the prime factors of F_5 are of the form $k \times 2^{5+1} + 1 = 64k + 1$.

However, using Lucas' version of the theorem will reduce the number of possibilities by a factor of one half. This becomes $128k + 1$. So, we have only to check 129, 257, 385, 513, 641, 769, 897, ... Among those numbers, only 257 and 641 are prime. Or, 257 is the value of F_3, and we just proved that every pair of Fermat Numbers are co-prime. Thus, 257 is eliminated. Brilliantly, Euler noticed that $641 - 1 = 2^7 \times 5$ and also that $2^4 = 641 - 5^4$.

For the sake of simplicity and clarity, we will use modular arithmetic to illustrate this proof.

Based on Euler's observation, $2^7 \times 5 \equiv -1 \pmod{641}$.

By raising both sides to the fourth power, we get

$$(2^7 \times 5)^4 \equiv 2^{28} \times 5^4 \equiv (-1)^4 \equiv 1 \pmod{641}$$

On the other hand,

$$2^4 \equiv 641 - 5^4 \equiv -5^4 \pmod{641}.$$

Therefore, by plugging the latter result into the former, we get

$$2^{28} \times 5^4 \equiv 2^{28} \times (-2^4) \equiv 1 \pmod{641}$$

Equivalently, $2^{32} \equiv -1 \pmod{641}$. Then, $2^{32} + 1 \equiv 0 \pmod{641}$. And thus, 641 divides $F_5 = 2^{32} + 1$, so, F_5 is composite.

■

Computing and factoring Fermat Numbers were laborious tasks since they require a lot of calculations. Nevertheless, they have become easier and faster due to the revolutionary development of computers. As a matter of fact, many distributed computing projects were established in which volunteers can participate by installing, on their computers, a program designed to use their CPU's power to factorize Fermat Numbers. Similarly to GIMPS Project that hunts for Mersenne Primes (as well as other kinds of primes), Fermat Search is a project dedicated to searching for Fermat Numbers divisors. [63]

The following is a list of some distributed computing projects and their websites. I encourage you to check them because they are a good option for those who want to learn more not only about Fermat numbers, but also about other kinds of numbers in general, and primes in particular.

You can also join one or more of those projects. They are free.

- Fermat Search: www.fermatsearch.org
- PrimeGrid: www.primegrid.com
- Proth Search: www.prothsearch.com
- GIMPS: www.mersenne.org
- Distributed.net: www.distributed.net

As of August 2020, 309 Fermat Numbers are proved to be composite and in total, 353 prime factors of them are known, 335 among them were found using the computer. [64]

It is not known whether there exist Fermat Primes other than the first four ones. However, English mathematician **John Conway** and **Kent Boklan**, a professor of Computer Science from CUNY Queens College, published a scientific paper in 2016, in which they predicted that the probability of the existence of another Fermat Prime is less than one billionth. [65]

Do date, the last prime factor of a Fermat Number, discovered on June 30th, 2020 by Gary Gostin and Fermat Search, is

$$171\,369\,935 \times 2^{11\,077} + 1$$

and it is a factor of $F_{11\,075}$. [64]

The following are the first 11 Fermat Numbers completely factored (they are the only completely factored Fermat Numbers):

Pierre de Fermat

F_n	$2^n + 1$	Status and decomposition if possible
F_0	$2^1 + 1$	3 is prime
F_1	$2^2 + 1$	5 is prime
F_2	$2^4 + 1$	17 is prime
F_3	$2^8 + 1$	257 is prime
F_4	$2^{16} + 1$	65,537 is the largest known Fermat prime
F_5	$2^{32} + 1$	= 4,294,967,297 =641 × 6,700,417 (fully factored in 1732 by Euler)
F_6	$2^{64} + 1$	= 18,446,744,073,709,551,617 =274,177 × 67,280,421,310,721
F_7	$2^{128} + 1$	=340,282,366,920,938,463,463,374,607,431,768,211,457 = 59,649,589,127,497,217 × 5,704,689,200,685,129,054,721
F_8	$2^{256} + 1$	=115,792,089,237,316,195,423,570,985,008,687,907,853,2 69,984,665,640,564,039,457,584,007,913,129,639,937 = 1,238,926,361,552,897 × 93,461,639,715,357,977,769,163,558,199,606,896,584,051, 237,541,638,188,580,280,321
F_9	$2^{512} + 1$	=13,407,807,929,942,597,099,574,024,998,205,846,127,47 9,365,820,592,393,377,723,561,443,721,764,030,073,546,9 76,801,874,298,166,903,427,690,031,858,186,486,050,853, 753,882,811,946,569,946, 433,649,006,084,097 =2,424,833 × 7,455,602,825,647,884,208,337,395,736,200,454,918,783,3 66,342,657 × 741,640,062,627,530,801,524,787,141,901,937,474,059,94 0,781,097,519,023,905,821,316,144,415,759,504,705,008,0 92,818,711,693,940,737
F_{10}	$2^{1024} + 1$	=179,769,313,486,231,590,772,930...304,835,356,329,624, 224,137,217 = 45,592,577 × 6,487,031,809 × 4,659,775,785,220,018,543,264,560,743,076,778,192,897 × 130,439,874,405,488,189,727,484...806,217,820,753,127,0 14,424,577
F_{11}	$2^{2048} + 1$	=32,317,006,071,311,007,300,714,8...193,555,853,611,059, 596,230,657 (617 digits) = 319,489 × 974,849 × 167,988,556,341,760,475,137 × 3,560,841,906,445,833,920,513 × 173,462,447,179,147,555,430,258...491,382,441,723,306,5 98,834,177 (564 digits)

[66]

Applications of Fermat Numbers

In his *Disquisitiones Arithmeticae*, **Carl Friedrich Gauss** revealed a link between Fermat Numbers and the constructibility of regular polygons.

He stated what he claimed, without proof, however, to be a sufficient condition for a regular polygon to be circumscribed about a circle, in other words, to be constructed using a geometer's toolkit, a straightedge and a compass.

This theorem, proved in 1836 by French mathematician **Pierre Wantzel** and called since then Gauss-Wantzel Theorem, states that:

An n-gon (an n-sided polygon) is constructed with a straightedge and a compass if and only if n is a product of a power of 2 and distinct Fermat Primes, i.e., $n = 2^k \times F_{i_1} \times F_{i_2} \times ... \times F_{i_n}$, where F_{i_k} are distinct Fermat Primes. [66+67+68]

Based on that, an n-sided regular polygon is constructible under the conditions stated above, if n = 3, 4, 5, 6, 8, 12, ... (A003401 in OEIS).

However, an n-sided regular polygon cannot be constructed using only the ruler and the compass if n = 7, 9, 11, 13, 14, 18, 19, 21, 22, 23, ... (A004169 in OEIS).

Another application of Fermat Numbers, more specifically Fermat primes, is in generating pseudo-random numbers similarly to Mersenne Twister. However, explaining how this works, is far beyond the scope of this book.

Generalized Fermat Numbers

Generalized Fermat Numbers are numbers of the form $a^{2^n} + b^{2^n}$, where a and b are positive non-zero coprime integers. [69]

Particularly, Generalized Fermat Numbers of the form $a^{2^n} + 1$ are often denoted by $F_n(a)$. [69]

Notice that a Fermat Number is equivalent to $F_n(2)$.

For example, $F_4(5) = 5^{2^4} + 1 = 152{,}587{,}890{,}626$.

If $F_n(a)$ is prime, then it is called a **Fermat Prime to base a.** Note that **if $F_n(a)$ is prime, then a must be odd**, otherwise it will be divisible by 2.

As of August 2020, the largest Generalized Fermat Prime was discovered in November 2018 and it has 6,317,602 digits which may require only 2.744×6 ! pages to print it.

(Note that "!" means factorial, and the number stated above is approximately 1975 assuming there are 80 digits per line, 40 lines per page).

To date, this number which is equal to $1{,}059{,}094^{1{,}048{,}576} + 1$, is ranked 14 among the biggest known prime numbers. [70]

It is still unknown if there exist infinitely many Generalized Fermat Primes.

The following are some generalized Fermat Numbers to some bases:

a	Generalized Fermat numbers in base a	OEIS
2	3, 5, 17, 257, 65537, 4294967297, ...	A000215
3	4, 10, 82, 6562, 43046722, ...	A059919
4	5, 17, 257, 65537, 4294967297, 18446744073709551617, ...	A000215
5	6, 26, 626, 390626, 152587890626, ...	A078303
6	7, 37, 1297, 1679617, 2821109907457, ...	A078304

Chapter 11

The Riemann Hypothesis

Bernhard Riemann

(1826 - 1866)

> *One would of course like to have a rigorous proof of this, but I have put aside the search for such a proof after some fleeting vain attempts because it is not necessary for the immediate objective...*
>
> **Bernhard Riemann***

*John Derbyshire, *Prime Obsession: Bernhard Riemann and the Greatest Unsolved Problem in Mathematics* (2003)

If you are interested in prime numbers or number theory in general, I am almost sure that you heard, at least once, about the famous Riemann Hypothesis and maybe you are already familiar with the Riemann's zeta(ζ) function.

The whole story began in Germany, in 1859, when a new elected member presened to The Berlin Academy, a brief scientific paper titeled *"Ueber die Anzahl der Primzahlen unter einer gegebenen Grösse"*, in English: *"On the number of primes less than a given magnitude"*. **[1]**

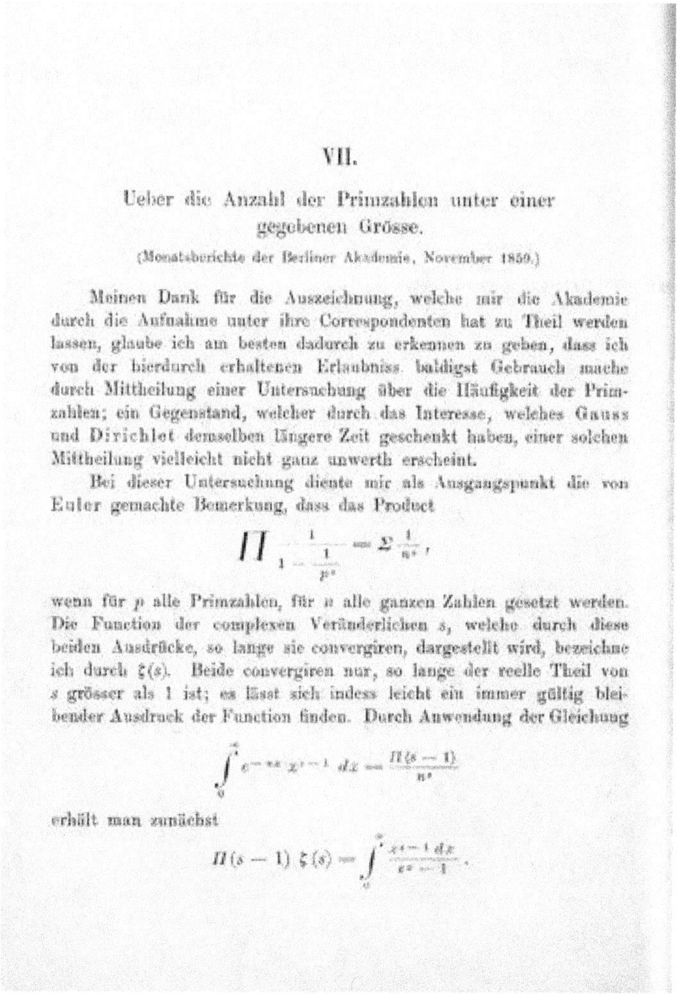

VII.

Ueber die Anzahl der Primzahlen unter einer gegebenen Grösse.

(Monatsberichte der Berliner Akademie, November 1859.)

Meinen Dank für die Auszeichnung, welche mir die Akademie durch die Aufnahme unter ihre Correspondenten hat zu Theil werden lassen, glaube ich am besten dadurch zu erkennen zu geben, dass ich von der hierdurch erhaltenen Erlaubniss baldigst Gebrauch mache durch Mittheilung einer Untersuchung über die Häufigkeit der Primzahlen; ein Gegenstand, welcher durch das Interesse, welches Gauss und Dirichlet demselben längere Zeit geschenkt haben, einer solchen Mittheilung vielleicht nicht ganz unwerth erscheint.

Bei dieser Untersuchung diente mir als Ausgangspunkt die von Euler gemachte Bemerkung, dass das Product

$$\prod \frac{1}{1-\frac{1}{p^s}} = \sum \frac{1}{n^s},$$

wenn für p alle Primzahlen, für n alle ganzen Zahlen gesetzt werden. Die Function der complexen Veränderlichen s, welche durch diese beiden Ausdrücke, so lange sie convergiren, dargestellt wird, bezeichne ich durch $\zeta(s)$. Beide convergiren nur, so lange der reelle Theil von s grösser als 1 ist; es lässt sich indess leicht ein immer gültig bleibender Ausdruck der Function finden. Durch Anwendung der Gleichung

$$\int_0^\infty e^{-nx} x^{s-1}\, dx = \frac{\Pi(s-1)}{n^s}$$

erhält man zunächst

$$\Pi(s-1)\,\zeta(s) = \int_0^\infty \frac{x^{s-1}\, dx}{e^x - 1}.$$

Figure 1: The first page of Bernhard Riemann's article concerning the number of primes that are less than a given magnitude.

The Riemann Hypothesis

This 8-page paper is now considered as a revolution in number theory because it explained, in a new way, the growth pattern of prime number as well as of composite numbers. That day, **Bernhard Riemann** (1826-1866) surprised mathematicians with a beautiful formula that could explain the distribution of prime numbers. [2]

One of the mysteries of primes is the irregularities of their growth. In other words, we cannot predict, with accuracy, where the next prime will be. If you take a look at a list of ordered prime numbers under 1000 for example, you will notice that the gap between them is not constant nor does it obey some kind of sequences. Sometimes, you find the primes clustered, which is the case with 2,3,5,7, but you may also notice that the gap between two consecutive numbers can be relatively bigger, which is the case of 887 and 907.

List of prime numbers under 1000:

2	3	5	7	11	13	17	19	23	29	31	37
41	43	47	53	59	61	67	71	73	79	83	89
97	101	103	107	109	113	127	131	137	139	149	151
157	163	167	173	179	181	191	193	197	199	211	223
227	229	233	239	241	251	257	263	269	271	277	281
283	293	307	311	313	317	331	337	347	349	353	359
367	373	379	383	389	397	401	409	419	421	431	433
439	443	449	457	461	463	467	479	487	491	499	503
509	521	523	541	547	557	563	569	571	577	587	593
599	601	607	613	617	619	631	641	643	647	653	659
661	673	677	683	691	701	709	719	727	733	739	743
751	757	761	769	773	787	797	809	811	821	823	827
829	839	853	857	859	863	877	881	883	887	907	911
919	929	937	941	947	953	967	971	977	983	991	997

[3.1]

List of the gaps between each two consecutive prime numbers under 1000

1	2	2	4	2	4	2	4	6	2	6	4
2	4	6	6	2	6	4	2	6	4	6	8
4	2	4	2	4	14	4	6	2	10	2	6
6	4	6	6	2	10	2	4	2	12	12	4
2	4	6	2	10	6	6	6	2	6	4	2
10	14	4	2	4	14	6	10	2	4	6	8
6	6	4	6	8	4	8	10	2	10	2	6
4	6	8	4	2	4	12	8	4	8	4	6
12	2	18	6	10	6	6	2	6	10	6	6
2	6	6	4	2	12	10	2	4	6	6	2
12	4	6	8	10	8	10	8	6	6	4	8
6	4	8	4	14	10	12	2	10	2	4	2
10	14	4	2	4	14	4	2	4	20	4	8
10	8	4	6	6	14	4	6	6	8	6	

[3.2]

Note that in this example, for the sake of simplicity, we talked about small prime numbers (under 1000). However, the gaps between prime numbers are usually studied as the primes get bigger and bigger.

Today, in our hands lies the beautiful result of Riemann's work. However, we do not have any record of how he got his astonishing ideas, except for some notes that give a blurry image of the steps he went through. It would be a wonderful experience to take a walk in Riemann's mind while he was disappointed by a dead-end and excited by a fruitful result. Nevertheless, if you are interested to read more about Riemann's work, many mathematicians such as **Carl Ludwig Siegel** (1886-1981) attempted to reconstruct, based on Riemann's notes and some other resources, the way this great mind got to his breathtaking achievement. [4]

Born on December 17, 1826, in Breselenz, Hanover, Germany, and died on July 20, 1866, in Selasca, Italy, German mathematician **Georg Friedrich Bernhard Riemann,** remains one of the greatest minds that have shaped our understanding of many fields of mathematical researches. [5]

Being the second among six children in a poor family did not prevent Riemann from attending the middle school, the high school, and finally the University of Göttingen. [5]

From a young age, Riemann had been interested in science, especially in mathematics. In fact, he devoured **Adrien-Marie *Legendre*'s** two-volume book *Number Theory* (1830) in just one week.[5] This book was for him a great introduction to number theory and it may be the reason behind his interest in the distribution of prime numbers since it contained Legendre's latest work on the (asymptotic) growth of primes.

Euler's *Introduction to Infinitesimal Analysis* was also one of the books that shaped Riemann's mathematical skills and had a great influence on his calculus-based approach to describe "the behavior" of prime numbers. [6]

When he went to the University of Göttingen, Riemann was planning to study Christian Theology and become a pastor to support his family. However, he ended up persueing his passion for studying mathematics.

A few years later, Riemann was nominated as a professor of mathematics at the University of Göttingen, following in his professor **Carl Friedrich Gauss'** and the famous mathematician **Johann Peter Gustav Lejeune Dirichlet**'s footsteps. [5]

Even though Riemann's life was shortly cut by a Tuberculosis when he was almost 40, he has made, during the few years he lived, an extraordinary quality work, not just in number theory, but also in many other branches of mathematics such as complex and real analysis, geometry, ... [5]

Before his revolutionary contribution to number theory, Riemann was best known for his work in geometry. In fact, he developed a kind of geometry different from the Euclidean geometry that we learned at school, and it was named after him, the Riemannian geometry. While the Euclidean geometry deals with planes, Riemannian geometry copes with curves and curved spaces such as the surface of a sphere, ... (Imagine doing geometry on the surface of an apple! Interesting!). This work was genuinely astonishing since it broadened our horizons and changed the way we think of geometry.

[6(p66-69)]

In his book *Stalking the Riemann Hypothesis*, **Dan Rockmore** affirms that *"Riemannian geometry quantifies the local variation of the real world ... [and] makes sense of a topography that includes craggy peaks, smooth dales, rolling hills, and flat deserts "*.

Riemannian geometry was indeed the mathematical basis for the well-known physicist **Albert Einstein'**s famous Theory of Relativity (General Relativity specifically). Einstein used the Riemannian geometry to describe the curvature of spacetime, to explain the nature of gravity, ... Even though an advanced level of mathematics is required to understand Einstein's application to Riemannian geometry, enjoying simplified explanations of what it is about is truly interesting. [5+7]

Riemann's work in number theory

Since that day at The Berlin Academy, Riemann's paper on the distribution of prime numbers, which is his only contribution to number theory, has been considered as a milestone in mathematics, since it used a different approach to tackle this problem.

Riemann is one of the first mathematicians to use analytic methods including calculus when dealing with problems in number theory. Not only did he use calculus but also, believe it or not, complex numbers. To be more specific, Riemann used what is known today as complex analysis, a branch of mathematics that deals with functions with complex variables, and that was first developed by Riemann and French mathematician **Augustin Louis Cauchy.** [6(p69) +8]

Nowadays, complex analysis has many real-world applications in engineering and also in many branches of physics such as quantum mechanics. [9]

To understand, superficially at least, what Riemann's Hypothesis is about, we should first discuss what complex numbers are.

Complex numbers

Numbers are known since antiquity. Ancient civilizations studied numbers because they used them, essentially, for counting days, months, and years, etc. However, over the centuries, our understanding of what numbers are has become deeper and we started to look at numbers from an abstract perspective. The Greeks were among the first civilizations to study the properties of numbers mathematically. For instance, one of the first and most prominent mathematicians in history is the Greek **Euclid.**

Natural numbers were the first to be discovered which is obvious since they were necessary for counting. Then, we needed to calculate the ratio of two numbers, especially when dealing with division (e.g. dividing an amount of money among a given number of people, ...). So, rational numbers such as $\frac{1}{3}, \frac{2}{7}, \frac{11}{19}, \ldots$ were invented. Afterward, the Greeks discovered a new kind of numbers that was somehow strange, as they thought.

To illustrate this, make a square with a side length of 1 unit. What is the length of its diagonal? Well, a primary school pupil would be able to answer this question by just applying the Pythagorean Theorem. The answer is $\sqrt{2}$. At first, they thought this was a rational number. However, it turned out that it is not.

It is really easy to prove this fact by way of contradiction.

Suppose that $\sqrt{2}$ is rational. Then, it can be written as the quotient $\frac{p}{q}$, where p and q are natural co-prime numbers (they have to be co-prime so that $\frac{p}{q}$ cannot be further reduced, i.e., it is an irreducible fraction). Thus, $\sqrt{2} = \frac{p}{q}$. Squaring both sides, we get $2 = \frac{p^2}{q^2}$. Hence, $2 \times q^2 = p^2$. Therefore, p^2 is even. The only way this is possible is that p itself is even, which implies that p = 2k, where k is a natural number. Thus, p^2 must be divisible by 4. So q has to be divisible by 2. This is absurd since it contradicts the fact that p and q are co-prime. Therefore, our initial hypothesis is false and hence, $\sqrt{2}$ is not rational. So, it is irrational. We can also prove easily that $\sqrt{3}, \sqrt{5}, \ldots$ are irrational. Not only this kind of numbers was considered as irrational, but also many others such as π. What is special about this number is that you cannot construct it using only the straightedge and the compass. I am not judging you, dear reader, but indeed no one can. Irrational numbers with this property are called transcendental numbers. Those numbers cannot be expressed using simple operators such as +, -, ×, /, $\sqrt[n]{}$ (nth root), ...

While numbers that cannot be constructed are called transcendental, those that can be constructed using the ancient Greek rules are called algebraic numbers. For example, rational numbers are algebraic and they are easy to construct. The same goes for irrational numbers that are the root of a positive integer and which can be constructed using the Pythagorean Theorem.

[6(pp71-73)]

There is indeed a beautiful method to get, geometrically, the square root of any integer by constructing a spiral (called the spiral of Theodorus) consisting of right triangles. It is based on the fact that $\sqrt{n}$ represents the hypothenuse of a right triangle with $\sqrt{n-1}$ and 1 are the lengths of the other two sides.

[10]

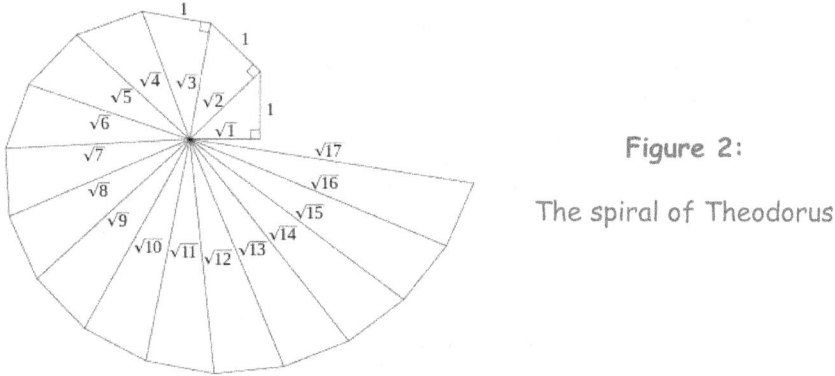

Figure 2:

The spiral of Theodorus

There is another method that is straightforward and doesn't require getting through all the other squares. It is based on a simple relation in a right triangle and it is illustrated in the following figure.

By taking one of a or b to be 1 and the other one to be a positive integer, then you will get the square root of that integer.

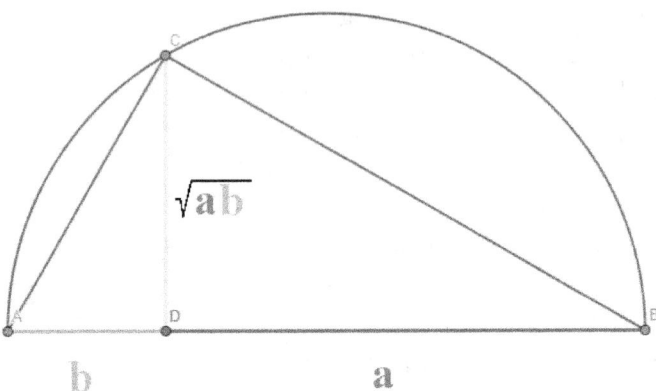

Figure 3: A metric relation in a right triangle

It is important to mention that natural numbers form the set ℕ. This is just a notation to used instead of saying the set of natural numbers. Then, comes ℤ, the set of whole numbers (also known as the set of integers). ℤ contains all the positive integers which are the natural numbers as well as all their opposites which are all the negative integers. This set is, obviously, bigger than ℕ, since the former part or, more echnically, a subset of the latter. Then comes the set ℚ that consists of all the rational numbers as well as all the integers (ℤ). The set ℚ, that englobes ℤ, and thus ℕ, forms with the set of irrational numbers, the set of real numbers ℝ. The following diagram may give you an intuition of how this works.

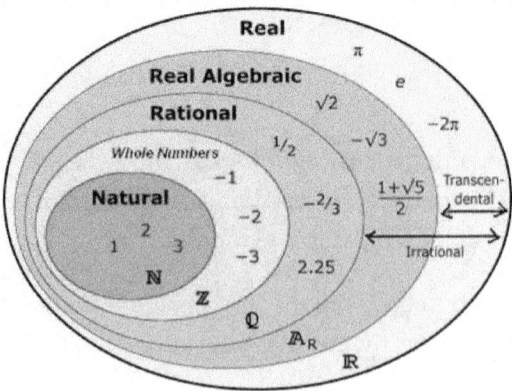

Figure 4: Different sets of numbers

> Note that until that point, 0 and negative numbers were considered as impossible. However, sets of different kinds of numbers have evolved to contain them, as well as other kinds of numbers.

However, here is a slightly different kind of numbers. What about $\sqrt{-1}$?

If you haven't taken a course about complex numbers yet, you might be wondering how this can be possible.

Well, it is worth mentioning that the way we see numbers has changed a lot. Numbers are seen as solutions to equations. For example, what is the number x such that 3x-2 = 0? It's $\frac{2}{3}$. Here is another one. Find x such that $x^2 - 1 = 0$. This is easy. The answer is ± 1. Another example is $x^2 + 1 = 0$. Well, this may not as easy as the previous ones. If it existed, the answer to this one should be $\sqrt{-1}$, and it is indeed. It seems we are in the same situation the Greeks faced when they discovered $\sqrt{2}$. $\sqrt{-1}$ is not a real number (it is neither an integer, a rational, nor an irrational number). Well, it looks like this is a new kind of numbers. And it is indeed. It is called an imaginary number. **[6(p72+p23)+11]**

More specifically, $\sqrt{-1}$, often denoted by *i,* is called the imaginary unit. Imaginary numbers are of the form $k \times i,$ where k is a real number.

There is another kind of numbers that is more "complex" than imaginary numbers, and those numbers are called, ironically, complex numbers. A complex number is formed of two parts: a real part and an imaginary part. In other words, a complex number is of the form $a + ib$, where a and b are real numbers. As the name indicates, the real part is a and the imaginary part is the number attached to *i* which is b. Please be careful that the imaginary part is just *b* and not *ib*.

In general, numbers can be thought of as solutions to some sort of equations, and complex numbers are not an exception.

As a little assignment and to make sure you grasp this concept well, solve the following equation, and determine the real part and the imaginary part of each solution, $(x + 1)^2 + 4 = 0$.

Let's solve it. (The symbol ⇔ means "implies")

$(x + 1)^2 + 4 = 0$

⇔ $(x + 1)^2 = -4$

⇔ $x + 1 = \pm\sqrt{-4}$, $\sqrt{-4}$ can be written as $i\sqrt{4} = 2i$ since $\sqrt{-4} = \sqrt{4}\sqrt{-1}$

⇔ $x = -1 \pm 2i$.

Thus, we have two solutions which are $-1 - 2i$ and $-1 + 2i$. The real part is the same for the two solutions and it is -1. However, the imaginary part of the first solution is -2, and it is 2 (or +2) for the second one.

Complex numbers are contained in the set of complex numbers ℂ.

ℂ englobes numbers of the form $a + ib$, the set of purely imaginary numbers (of the form $k \times i$) as well as ℝ, the set of real numbers. Based on this fact, any real number can be considered as a complex number with an imaginary part being 0.

While a number line is used to represent real numbers, a complex plane is a good way of graphing complex numbers. A complex plane is composed of two perpendicular number lines: a horizontal axis called the real axis, and a vertical one named, the imaginary axis, which are equivalent, relatively, to the x-axis and the y-axis in the Cartesian plane. Any complex number can be represented as a point on the complex plane defined by a unique pair of ordered numbers which are the real and the imaginary part of the complex number in question. [12]

Here is an illustration of some complex numbers represented as points on the complex plane:

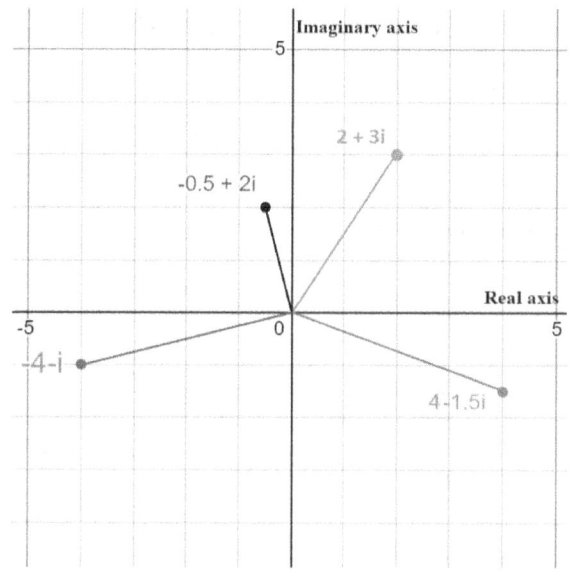

Figure 5:

Graphical representation of some complex numbers

The name "imaginary" appeared for the first time in the 17th century, in **Descartes**'s *La Géometrie* and it was meant to be disparaging since, at that time, mathematicians thought that imaginary numbers, as well as negative numbers, were useless. However, it turned out that they are extremely important. [13+14]

Cardano (1501 –1576) and **Rafael Bombelli** (? – 1572) were among the first to acknowledge the existence of complex numbers. In fact, Cardano even used them when he was working on a generalized solution to cubic equations. However, at that point, mathematicians' understanding of what complex numbers are was rudimentary. Complex numbers started to be deeply understood when many mathematicians such as **Euler, Abraham de Moivre, Augustin Louis Cauchy, Bernhard Riemann**, ... developed the theory of complex analysis, which became an independent branch of mathematics. [14(p 121-123)]

By now, you may have at least a little understanding of what complex numbers are and how they can be represented in the complex plane. This will help you understand some details of the Riemann Hypothesis.

128   The Riemann Hypothesis

In his theory, Riemann introduced his famous Zeta (ζ) function which is the following:

$$\zeta(s) = 1 + \frac{1}{2^s} + \frac{1}{3^s} + \frac{1}{4^s} + \frac{1}{5^s} + \frac{1}{6^s} + \cdots$$

(The dots "..." mean "continuing indefinitely")

In a compact form,

$$\zeta(s) = \sum_{k=0}^{\infty} \frac{1}{s^k}$$

where s is a complex number

(Σ is the Sigma summation notation) [15]

> Note that the Riemann zeta function may seem like a simple sum of the trms of a geometric sequence, but it is a sum of an infinite number of terms. This kind of sums is called an infinite series.

Historically, Riemann was not the first to study the zeta function.

A particular case of it, when s = 1, i.e., the infinite series of the reciprocals of natural number, was first studied in the 14th century by French mathematician **Nicholas Oresme** (c. 1325 – 1382) who proved that $\zeta(1) = 1 + \frac{1}{2} + \frac{1}{3} + \frac{1}{4} + \frac{1}{5} + \frac{1}{6} + \cdots$ is divergent which means that this sum is not finite, it is indeed infinite. This particular infinite sum is called the Harmonic series and it is truly an interesting topic to learn about. The name is related to the concept of Harmonics in music, which are overtones higher than the fundamental frequency of a tone, but this is a story for another time. However, I am almost sure that you will be surprised by the applications of the Harmonic series, especially in puzzles such as the "worm on the rubber band problem", the "block-stacking problem",...   [16+17]

In 1737, in a paper entitled *Variae observationes circa series infinitas*, **Euler**, then at the age of 28, solved the Basel Problem, suggested in 1650 by **Pietro Mengoli**, which consisted of calculating the following infinite series

$$1 + \frac{1}{2^2} + \frac{1}{3^2} + \frac{1}{4^2} + \frac{1}{5^2} + \frac{1}{6^2} + \cdots$$

i.e.,

$$\zeta(2) = \sum_{k=0}^{\infty} \frac{1}{2^k}$$

[18+19]

This problem challenged many mathematicians until Euler came out with the correct answer which is $\frac{\pi^2}{6}$. Intuitive, right? You may be wondering why π appears in such an infinite series that, apparently, doesn't have anything to do with circles and curves. The proof is a little bit tricky, so it won't be included in this book. However, I encourage you to read about it on your own.

Euler didn't prove just that $\zeta(2) = \frac{\pi^2}{6}$ but also that

$$\zeta(4) = \sum_{k=0}^{\infty} \frac{1}{4^k} = \frac{\pi^4}{90}$$

$$\zeta(6) = \sum_{k=0}^{\infty} \frac{1}{6^k} = \frac{\pi^6}{945}$$

$$\zeta(26) = \sum_{k=0}^{\infty} \frac{1}{26^k} = \frac{2^{24}}{27! \times 76{,}977{,}927 \times \pi^{26}}$$

$$\zeta(26) = \sum_{k=0}^{\infty} \frac{1}{26^k} = \frac{2^{24}}{27! \times 76{,}977{,}927 \times \pi^{26}}$$

$$1 + \frac{1}{3^2} + \frac{1}{5^2} + \frac{1}{7^2} + \frac{1}{9^2} + \cdots = \frac{\pi^2}{8}$$

[20]

130 The Riemann Hypothesis

After solving all those particular cases, Euler, as he did usually, studied the general case, i.e.,

$\zeta(n) = 1 + \frac{1}{2^n} + \frac{1}{3^n} + \frac{1}{4^n} + \frac{1}{5^n} + \frac{1}{6^n} + \cdots$, where n is an integer.

Euler managed to prove, using some tricky techniques, that this infinite series is equal to an infinite product, a special product, indeed.

He showed that,

$$\zeta(n) = \frac{1}{1-\frac{1}{2^n}} \times \frac{1}{1-\frac{1}{3^n}} \times \frac{1}{1-\frac{1}{5^n}} \times \frac{1}{1-\frac{1}{7^n}} \times \frac{1}{1-\frac{1}{11^n}} \times \frac{1}{1-\frac{1}{13^n}} \times \cdots$$

As you may have noticed, it is equal to the product of an infinite number of terms of the form $\frac{1}{1-\frac{1}{p^n}}$, where p ranges over the set of positive prime numbers.

In a compact form, this may be written, using the Pi product notation, as

$$\zeta(n) = \prod_{p=2}^{\infty} \frac{1}{1-\frac{1}{p^n}}$$

where p is a prime number. [18+19]

If you are not surprised yet, well, you should be because this is a golden connection between a sum involving all positive integers and a product involving only prime numbers.

Euler's zeta function dealt just with real numbers, particularly with integers. However, in Riemann's zeta function, the parameter s is a complex number. What Riemann did was not just changing some letters, rather he studied the (analytic) properties of this special function since it has a strong relationship with prime numbers. He searched for the zeros of $\zeta(s)$, i.e., of the numbers s such that $\zeta(s) = 0$. Note that if s = 1, then we get the Harmonic series we discussed earlier. However, if n > 1, then the series converges to a definite number different from 0.

A certain kind of roots of this function is called trivial zeros. They are -2, -4, -6, -8, ... More generally, $\zeta(-2n) = 0$, where n is a positive non-zero integer. They are called "trivial" because proving they are roots of the equation stated above is relatively easy. You may be wondering why this is true. Explaining this may require advanced knowledge of calculus and complex analysis. However, this may seem obvious to you if you know that the Riemann's zeta function may be written as:

$$\zeta(s) = 2^s \pi^{s-1} \sin\left(\frac{\pi s}{2}\right) \Gamma(1-s) \zeta(1-s)$$

[21]

Without getting in many details, if s = -2n, where n is a positive integer, then $\sin\left(\frac{\pi s}{2}\right) = \sin\left(\frac{-2n\pi}{2}\right) = \sin\left(-n\frac{\pi}{2}\right) = 0$.

In fact, even if s = 2n, then the sine function will be zero, but in this case, this expression will not be defined.

If you did not understand the previous argument, don't worry, you don't have to. I included it just to give you, dear reader, a short superficial explanation of why this is true.

In addition to the trivial zeros, the Riemann zeta function has another kind of roots called the non-trivial zeros because it is not easy to prove them.

It is known that all the non-trivial zeros are complex numbers that have a real part between 0 and 1, more technically, a non-trivial zero lies in the open strip $\{s \in \mathbb{C}: 0 < \text{Re}(s) < 1\}$. This strip is called the critical strip.

Bernhard Riemann claimed, in his paper, that all the non-trivial zeros, which are infinite as he conjectured, are complex numbers with a real part being $\frac{1}{2}$, in other words, the non-trivial zeros are complex numbers of the form $\frac{1}{2} + ib$, where b is a real number. If you plot those solutions in the complex plane, you will get a vertical line perpendicular to the real axis, and obviously, parallel to the imaginary axis. This special line is called the critical line. Thus, the Riemann hypothesis states that all the non-trivial zeros lie on the critical line. [17]

132 The Riemann Hypothesis

To date, the Riemann hypothesis is one of the most important open problems in mathematics. Even though it is still a conjecture, mathematicians strongly believe that it is true. Indeed, numerically, up 10 trillion (10^{13}) non-trivial zero from the critical strip were verified by ZetaGrid* and all of them are located, as Riemann conjectured, on the critical line. [23+24]

> *ZetaGrid was a distributed computing project that was shut down in 2005 after almost 4 years of verifying around a billion non-trivial zeros of the Riemann Hypothesis daily. [25]

This numerical evidence suggests that this conjecture is experimentally true, hence, it can be used as a valid theorem in experimental fields such as physics. Speaking of which, the Riemann hypothesis has many applications in physics, especially, in quantum mechanics.

However, mathematically, regardless of what the numerical calculations may suggest, we cannot consider this hypothesis as a valid result unless a strong and flawless proof is discovered because, throughout history, many conjectures that were experimentally valid turned out to incorrect. For example, one of those conjectures deals with the prime-counting function $\pi(n)$ and the offset logarithmic integral function $Li(n)$ (see the prime-counting function $\pi(n)$). Gauss conjectured that $Li(n)$ is always greater than $\pi(n)$. This statement is true up to a very large number, around 1.397×10^{316}. [26]

If the Riemann hypothesis is true, which is very likely the case, then many important results in number theory will be proven easily. Besides, this will allow us to better understand the irregular behavior of prime numbers.

Even if Riemann's conjecture is false, mathematicians have realized many achievements in many relevant areas of mathematics such as number theory, and complex analysis, ... Those results have already enriched our knowledge of numbers.

Since 1859, Riemann's groundbreaking paper has been considered as a breakthrough in number theory, particularly, and in mathematics, generally.

This conjecture is extremely significant that the great German mathematician **David Hilbert** (1862 – 1943) included it as the 8th problem in his famous list consisting of 23 of the most important problems in mathematics. This list, published in 1902, is often called Hilbert's problems. [27]

For the last 160 years, many mathematicians have attempted to tackle this challenge. However, to date, no one has discovered a strong proof. For instance, in 1885, Dutch mathematician, **Thomas Joannes Stieltjes** claimed that he proved the Riemann hypothesis, but no such proof has never been published. It is very likely that even if his claim was right, his proof would contain a flaw. But what if his proof was right? [28]

One of the recent official attempts to prove Riemann's hypothesis was by English mathematician Dr. **Michael Atiyah**, who, unfortunately, passed away in 2019. [29] It is worth mentioning that Dr. Atiyah received the Fields Medal in 1966 and was awarded the Abel Prize in 2004 for his work in K-Theory, one of the interesting theories in topology. [30+31]

When he announced his discovery in September 2018, in the 2018 Heidelberg Laureate Forum, in Germany, he described his proof as "simple". He built his "simple" proof upon the work of two great mathematicians of the 20th century, **John von Neumann** and **Friedrich Ernst Peter Hirzebruch** (1927 – 2012). In simple words, Dr. Atiyah assumes Riemann's hypothesis does not hold, and he claimed to reach a logical contradiction, which means that R.H. must be true. However, many mathematicians are doubtful about this proof. [32+33+34]

After reading this far about Riemann's hypothesis, you may be interested in proving it and become one of the greatest mathematicians that history will never forget.

Have you finished the proof ? Yes ? That's great! Is there any flaw? No. Ok, congratulations! You have just won $ 1 million and an eternal fame.

As a matter of fact, on May, 24th, 2009, Riemann's Hypothesis was classified by the « Clay Mathematical Institute » (CMI)(a non-profit organization established in 1998 by Landon Clay whose primary objective *"is to encourage the increase and dissemination of mathematical knowledge".*) as one of the 7 Millenium Prize Problems, a list containing 7 amog the most important and the most challenging open problems in mathematics, and computer science. A $1 million will be awarded by this institute to the first discoverer(s) of a proof. [35]

Each year numerous papers related to Riemann's hypothesis are published. Who knows, maybe the next paper will be the proof we have been waiting for for more than 160 years.

Please note that this was just a very simplified introduction to Riemann's hypothesis. It would be so fascinating to dive into the details and be overwhelmed by the formulas and the theorems, but it is up to the curious reader to do so.

Chapter 12

Primes in Arithmetic Progressions

Have you heard before about arithmetic sequences? If yes, that's great. However, if not, you are missing something genuinely interesting.

A sequence is simply a list or a set of ordered numbers called the terms of the sequence. However, what is special about sequences is the connection that exists between its terms. There is always a pattern that the sequence follows. This pattern can be expressed as an expression used to calculate the nth term of the sequence.

An example of a sequence is

$$1, 2, 3, 4, 5, 6, 7, 8, \ldots$$

I am sure you are familiar with this one. This is just the sequence of natural numbers starting from 1. You may have noticed the simple pattern in this sequence: each term is one plus the previous one. Trivial, isn't it?

Here are some other examples,

$$6, 10, 14, 18, \ldots$$

$$1, 4, 7, 10, 13, \ldots$$

136   Primes in Arithmetic Progressions

The pattern in the two previous sequences is similar to the first one. However, each time, to get the next term, we add 4 to the previous one in the first example, and 3 in the second. In other words, we can say that the difference between two consecutive terms in the examples above is constant. A sequence with this property is called an arithmetic sequence, also an arithmetic progression. This is a famous kind of sequences that mathematicians have been studying for more than 2000 years.

More technically, a sequence, not necessarily arithmetic, is usually denoted by a small letter, (say a, b, ...) as (a_n)*. (a_n) is expressed in terms of n. For example, we can, succinctly, describe the sequence 6, 10, 14, 18, ... as $a_n = 4n + 2$, where n is an integer greater than or equal to one. The terms of this sequence (or any sequence in general) are indexed, which means that a_1 corresponds to the first term of the sequence which is equal to 4×1+2=6, a_2 to the second term, ...

If the initial term of an arithmetic progression is a_1 and the common difference between successive terms is d, then the nth term of the sequence (a_n) is given by

$$a_n = a_1 + d(n-1)$$

where n is a positive integer (it can be 0).

More generally, if a_i and a_j are two terms of the arithmetic sequence (a_n), then

$$a_i = a_j + d(i-j) \text{ (assuming } i \geq j)$$

*Please do not confuse (a_n) with a_n. The first denotes the sequence (a_n), however, the second stands for the nth term of the sequence (a_n).

Consider the following sequence: 2, 3, 5, 7, 11, 13, 17, 19, 23, ...

This is one of the most interesting sequences. It is the sequence of prime numbers.

Primes in Arithmetic Progressions

If you look at this sequence, it won't be easy to find a pattern like in the previous ones. This is because prime numbers behave irregularly. Nevertheless, mathematicians have spotted some signs of order in this apparent chaos. Surprisingly, sometimes primes follow a pattern, not for a long time, however. Sometimes, primes appear as consecutive terms of an arithmetic progression.

Primes in an arithmetic progression is a sequence of at least 3 consecutive prime numbers, i.e., the difference between them is constant. More technically, an arithmetic progression is a set of prime numbers of the form a + nb, where a and b are fixed positive integers, for consecutive integer values of n.

An arithmetic progression of k consecutive primes with a common difference a is sometimes denoted by AP-k or PAP-k. If the primes are of the form a+nb, then n can go from 0 to k-1.

For example, 3, 7, 11 is a sequence of prime numbers of the form 4k+3, where k =0, 1, and 2. This is an AP-3 with a common difference 4. Another AP-3 is the trivial 3, 5, 7 with a common difference 2. This is of the form 3+4n, n= 0, 1, 2. Can we find others ? Well, yes we can since there exist infinitely many triples of prime numbers in arithmetic progression. This important result was proved in 1939 by Dutch mathematician **Johannes Gaultherus van der Corput** (1890 –1975). [1]

British mathematician **Roger Heath-Brown** proved that there exist infinitely many arithmetic progressions of 4 terms where 3 are primes and one is either a prime or a semiprime (the product of two prime numbers). [2]

---

*A semiprime

It is also called a 2-almost prime, or biprime is a composite number that is the product of two primes (not necessarily distinct). The first few are

4, 6, 9, 10, 14, 15, 21, 22, ... (OEIS A001358).

However, what about the general case ? Are infinitely many AP-5, AP-6, ... AP-1000,... ? This had been a conjecture since the 18th century, precisely since the 1770s when French mathematician **Joseph-Louis Lagrange** and British mathematician **Edward Waring** started to look at how large the common difference between primes in an arithmetic progression can be. [3]

Figure 1:
Edward Waring (1736 -1798)

In 2004, based on a recent work by the American mathematician **Daniel Goldston** and Turkish mathematician **Cem Yalçın Yıldırım**, as well as on a theorem that is related to primes in arithmetic sequences called Szemerédi's Theorem, Australian-American mathematician **Terence Chi-Shen Tao** and British mathematician and number theorist **Ben Joseph Green** proved the existence of an arbitrarily long arithmetic progressions of prime numbers.

Figure 3:
Terence Tao

Figure 3:
Ben Joseph Green

In other words, they proved that for any given positive integer n, there exists an AP-n. In the 56-page paper they published, this was stated as Theorem 1.1:

The prime numbers contain infinitely many arithmetic progressions of length k for all k.

[4]

Interestingly, it has been conjectured that the minimal possible difference in an AP-k is k#* for all k > 7. if k is a prime number and AP-k is an arithmetic progression of k prime numbers with a common difference a, and that starts with k, then a must be a multiple of $(k-1)$#.* [5+6]

> * While ''!'' denotes the factorial of a number, '' # '' denotes the primorial of a certain number, i.e. the product of all the prime numbers less than or equal to
>
> Let n be a positive integer.
>
> - n! is the product of all the positive integers less than or equal to n and greater than or equal to 1.
> $$n! = n \times (n-1) \times (n-2) \times (n-3) \times ... \times 3 \times 2 \times 1$$
> - n # is the product of all the prime numbers less than or equal to n.
> $$n! = p_n \times p_{n-1} \times p_{n-3} ... \times 5 \times 3 \times 2,$$
>   where p_n is the largest prime less than or equal to n.

For example, the AP-3 with primes {3, 5, 7} and a common difference 2# = 2, the AP-5 with primes {5, 11, 17, 23, 29} and a common difference 4# = 6, ...

The largest known example is AP-27, and it is, by the way, the first known of its kind, was discovered on September 23rd, 2019 by **Rob Gahan** and PrimeGrid, and it is

$$224,584,605,939,537,911 + 81,292,139 \cdot 23\# \cdot n, \text{ for } n = 0..26.$$

[6]

140 Primes in Arithmetic Progressions

Talking about records, the first AP-6 was discovered in 1967. It is

$$121{,}174{,}811 + 30k, k = 0,\ldots,5.$$

The first AP-7 was found almost 20 years later, in 1997.

In 1998, **H. Dubner, T. Forbes, N. Lygeros, M. Mizony, H. Nelson**, and **P. Zimmermann** discovered the first AP-10, and this was an incredible achievement. Indeed, in the 6-page paper they published, it was indicated that " ... *it is very likely that 10 primes will remain the record for a long time* ". [7]

This AP-10 is given by

100 996 972 469 714 247 637 786 655 587 969 840 329 509 324 689 190 04 1 803 603 417 758 904 341 703 348 882 159 067 229 719 + 210k.
(A033290 in OEIS)

6 years later, AP-23 was discovered by **Markus Frind, Paul Jobling**, and **Paul Underwood** which is given by

$$56{,}211{,}383{,}760{,}397 + 199{,}678 \cdot 23\# \cdot k \,, k = 0, \ldots, 22.$$

The first AP-24 was found in 2007 by **Jaroslaw Wroblewski** and it is of the form

$$468{,}395{,}662{,}504{,}823 + 205{,}619 \cdot 23\# \cdot k \,, k = 0, \ldots, 23.$$

A year later, AP-25 was discovered by **Raanan Chermoni** and **Jaroslaw Wroblewski**.

AP-25: $6171054912832631 + 366384 \cdot 23\# \cdot k, k = 0, \ldots, 24.$

AP-26 was discovered in 2010 by and it is of the form

$$43142746595714191 + 23681770 \cdot 23\# \cdot k \,, k = 0,\ldots, 25.$$

9 years later, AP-27 was found. [6]

Primes in Arithmetic Progressions

Additional information

The following table contains the AP-k with the smallest last term for the first few values of k: [8]

K	AP-k ($0 \leq n \leq k-1$)	Last term
3	$3 + 2n$	7
4	$5 + 6n$	23
5	$5 + 6n$	29
6	$7 + 30n$	157
7	$7 + 150n$	907
8	$199 + 210n$	1669
9	$199 + 210n$	1879
10	$199 + 210n$	2089
11	$110,437 + 13,860n$	249,037
12	$110,437 + 13,860n$	262,897
13	$4,943 + 60,060n$	725,663
14	$31,385,539 + 420,420n$	36,850,999
15	$115,453,391 + 4,144,140n$	173,471,351
16	$53,297,929 + 9,699,690n$	198,793,279
17	$3,430,751,869 + 87,297,210n$	4,827,507,229
18	$4,808,316,343 + 717,777,060n$	17,010,526,363
19	$8,297,644,387 + 4,180,566,390n$	83,547,839,407
20	$214,861,583,621 + 1,884,6497,670n$	572,945,039,351
21	$5,749,146,449,311 + 26,004,868,890n$	6,269,243,827,111

The following table contains the minimal common difference a_{min} of an AP-k for the first few values of k [9]

k	a_{min}	k	a_{min}
1	0	11	2310
2	1	12	2310
3	2	13	30030
4	6	14	30030
5	6	15	30030
6	30	16	30030
7	150	17	510510
8	210	18	510510
9	210	19	9669690
10	210	20	9699690

Chapter 13

Dirichlet's Theorem

One of the questions that are often asked about prime numbers is: Are there infinitely many primes of the form 6k+1, where k is a positive integer? This question can be asked differently as: Are there infinitely many primes in the sequence 1, 7, 13, 19, 25, 31, 37, 43, 49, 55, 61, ... ? Well, we can conjecture that there are, especially that there are many of them: 7, 13, 19, 31, ...

Here is a similar question: Are there infinitely many primes of the form 6k+2, i.e., are there infinitely many primes in the list: 2, 8, 14, 20, ... ?

This is obvious. The only prime is 2 and it is easy to prove this. 6k+2 can be written as 2(3k+1). So, if k > 0, this number is composite. The only possibility that it is prime is when k = 0.

This same argument can be applied to 6k+3, 4k+2, 8k+2, ...

You may have noticed that there is a common property between those numbers. They are all of the form a+kb, where a and b share a common factor (a, b, and k are natural numbers). In this case, you will find at most a prime number (when k = 0).

However, if a and b are co-prime, i.e., they do not share any common prime factor, then it seems that there are infinitely many prime numbers of the form a + kb, and there is, indeed.

Dirichlet's Theorem

This last statement is a theorem in number theory, called **Dirichlet's Theorem on Primes in Arithmetic Progressions**, and it states that:

If a and b are relatively prime positive integers, then the arithmetic progression a, a+b, a+2b, a+3b, ... contains infinitely many primes. [1]

In other words, this theorem states that there are infinitely many prime numbers of the form a+kb, where a and b are natural numbers, for consecutive integer values of k.

> Please do not confuse Dirichlet's theorem with AP, Dirichlet deals with the infinite number of primes in a+kb, wheras AP deals with primes in arithmetic progression.
>
> Dirichlet's Theorem doesn't say anything about consecutive primes in the sequence.

This theorem was first conjectured by the German mathematician **Carl Friedrich Gauss** in the late 18th–early 19th-century. In fact, in his *Disquisitiones Arithmeticae*, it was stated that:

> *The illustrious Le Gendre himself admits [that] the proof of the theorem — [namely, that] among [integers of] the form kt + l, [where] k and l are given coprime integers [and] t denotes a variable, surely prime numbers are contained — seems difficult enough, and incidentally, he points out a method that could perhaps lead to it; however, many preliminary and necessary investigations are [fore]seen by us before this [conjecture] may indeed reach the path to a rigorous proof.*

[2+3]

But it was first proved in 1826 by the German mathematician **Peter Gustav Lejeune Dirichlet** (1805 - 1859). [3]

Dirichlet's proof involves advanced methods such as calculus and other techniques from analytic number theory. As a matter of fact, he used what is known as Dirichlet's L-series that are an important tool in modern number theory. [4+5+6]

144 Dirichlet's Theorem

So far, we know that if a and b are positive co-prime integers, then the arithmetic progression a, a+b, a+2b, a+3b, ... contains infinitely many primes. However, we have no idea about when the first prime might occur.

Russian mathematician **Yuri Vladimirovich Linnik** (1915 – 1972) this matter. In 1944, he published two papers titled *On the least prime in an arithmetic progression* (*I* and *II*), in which he proved that a theorem called after him, **Linnik's theorem**, and that states that if we denote p(a, b) the least prime in the arithmetic progression a+kb, where k a positive integer and a and b are natural coprime numbers such that $1 \leq a < b$, then

$$p(a,b) < cb^L$$

for sufficiently large b, where L and c are positive integers. c is a constant and L is known as Linnik's constant. [7]

Linnik showed that L and c are effectively computable. However, he did not calculate them. Recent work has shown that $L \leq 5$. [8] Many mathematicians such as Polish mathematicians **Andrzej B. M.Schinzel** and **Wacław F. Sierpiński** have conjectured that $L = 2$. [9]

Dickson's conjecture: A Generalization of Dirichlet's Theorem

Dirichlet's proposition was generalized in 1904 by the American mathematician **Leonard Eugene Dickson** (1874 –1954) who is best known for his 3-volume *History of The Theory of Numbers*, considered as one of the most important books in number theory. [10]

While Dirichlet's Theorem assures that there exist infinitely many prime numbers of the form a+kb, for fixed a and b, Dickson conjectured that if there exist a sequence of linear expressions

$$a_1 + n \times b_1, \ a_2 + n \times b_2, \ a_3 + n \times b_3, \ \ldots, \ a_k + n \times b_k,$$

where a_i and b_i are integers such that $b_i > 1$, then there exist infinitely many positive integers n such all the $a_i + n \times b_i$ are prime numbers, simultaneously. [9+10]

However, this statement is not always true. In fact, in some cases such as 2n + 3, 2n + 5, 2n + 7, it doesn't hold because one of those numbers must always be divisible by 3: the first is divisible by 3 if 3 divides n, the second if 3 divides n + 1 and the third if 3 divides n−1.   [11]

More generally, the cases excluded are when there exists a prime number p that divides the product of all the $a_i + n \times b_i$ for all numbers n.   [9+12]

If this conjecture is true, then it will be used to prove multiple conjectures and open problems such that the Twin prime conjecture which states that there exist infinitely many pairs of primes of the form (p, p+2). The infinite number of Sophie Germain Primes and composite Mersenne Numbers, as well as the K-Tuple conjecture, will also be proved.   [9+11]

Hypothesis H: A Generalization of Dickson's conjecture

Dickson's conjecture, the generalization of Dirichlet's Theorem, was itself generalized.

Polish mathematicians **Wacław Franciszek Sierpiński** and **Andrzej Schinzel** generalized Dickson's conjecture into the Hypothesis H Conjecture. The linear expressions of the form $a_i + nb_i$ were substituted with irreducible polynomials* of the form

$$a_n x^n + a_{n-1} x^{n-1} + a_{n-2} x^{n-2} + \cdots + a_2 x^2 + a_1 x^1 + a_0,$$

where a_n is the leading coefficient, n is a positive integer, a_i are integers, and x is a variable(it is an integer in this case).

> *the gcd of all the coefficient is 1, i.e., there doesn't exist any common factor between all the coefficient.

146 Dirichlet's Theorem

The Hypothesis H states that

If $f_1(x), f_2(x), \ldots, f_k(x)$ (k is a positive integer) are irreducible polynomials with integral coefficients and positive leading coefficients, and for every prime number p, p doesn't divide the product

$$f_1(m) \times f_2(m) \times \ldots \times f_k(m),$$

i.e., p doesn't divide any of the polynomial values $f_i(m)$ for every integer m, where $1 \leq i \leq k$, then there exists a positive integer n such that $f_1(n), f_2(n), \ldots, f_k(n)$ are all primes. [13+14+15]

The second condition in the statement above must be respected; otherwise, there will be many counterexamples such as the 2 polynomials x+2 and x+5, for which n+2 and n+5 will never be primes simultaneously because, for any natural number n, one of them will always be even: if n is even, then the first is, too, and if it is odd, then the second is even. So, in both cases, their product will be even.

If true, this conjecture will solve an old question that Euler once asked. The question is whether there exist infinitely many prime numbers of the form $n^2 + 1$.

The first few of them are

n [16]	$n^2 + 1$ [17]
1	2
2	5
4	17
6	37
10	107
14	197
16	257
20	401
24	577
26	677
36	1297
40	1601
54	2917
56	3137
66	4357
74	5477
84	7057
90	8101
94	8837
110	12101

Though it may seem simple, this question appears as the 4th problem in Landau's four-problem list of unattackable open questions in number theory.

An important result concerning this problem was found by American-Polish mathematician **Henryk Iwaniec** who showed that there are infinitely many numbers of the form $n^2 + 1$ with at most two prime factors. However, whether there exist infinitely many numbers of this form with at most one prime factor seems to be beyond our reach at present. [18+19]

If you are interested in other conjectures involving primes and polynomials, you will be interested to learn about Bunyakovsky Conjecture, and also about Bateman–Horn Conjecture that is considered as a generalization of Hypothesis H. (So, we can say that Bateman-Horn conjecture is the generalization of the generalization of the generalization of Dirichlet's Theorem.) [20]

Chapter 14

Formulas for primes

This may sound shocking, but there exists a formula for the nth prime number. Indeed, there exist more than one.

One of those formulae is the function f defined over positive integers n by $f(n) = n$. Technically, this function generates infinitely many prime numbers and the reason is obvious since it yields all the natural numbers that contain all the prime numbers.

Dirichlet's Theorem can also be used to construct such formulae. For example, if we take a and b in the function $f(n) = an + b$ to be relatively prime positive integers, then this function will generate infinitely many primes.

However, those functions generate not only primes but also composite numbers.

Those formulae may not be interesting enough. What about a formula that generates only prime numbers?

Such formulas are functions $f(n)$ defined over positive integer n ≥ 1 satisfying one of the following conditions:

- $f(n) = p_n$, where p_n is the nth prime.
- $f(n)$ is always a prime number, and if $n \neq m$, then $f(n) \neq f(m)$
- the set of prime numbers is equal to the set of positive values assumed by the function.

In 1952, Polish mathematician **Wacław Sierpiński** discovered a method satisfying the first condition.

He showed that the if p_n is the nth prime number, then

$$p_n = \lfloor 10^{2^n} \alpha \rfloor - 10^{2^{n-1}} \lfloor 10^{2^{n-1}} \alpha \rfloor$$

where α is a constant defined by

$$\alpha = \sum_{i=1}^{\infty} \frac{p_i}{10^{2^i}} \approx 0.02030005000000070...$$

Notes

- $\lfloor x \rfloor$ denotes the floor function, i.e., it is the greatest integer less than or equal to x
- p_i is the ith prime number

[1]

A more generalized form of this function was given by English mathematicians **Godfrey Harold Hardy** and **Edward Maitland Wright** who suggested that if r is an integer greater than one, then

$$p_n = \lfloor r^{-n^2} \beta \rfloor - r^{2n-1} \lfloor r^{(n-1)^2} \beta \rfloor$$

where

$$\beta = \sum_{i=1}^{\infty} \frac{p_i}{r^{i^2}}$$

However, as you may have noticed, this formula is useless because to calculate p_n, one needs to know not only $p_1, p_2, p_3, ...,$ and p_{n-1} but also p_n itself as well as many other prime numbers. Unless we find another way to calculate the constant α (or equivalently β) without first calculating p_n, this function will have no practical application.

Another formula was included in **Paulo Ribenboim**'s *The new book of prime number records* and it states that

[1]

$$p_n = 1 + \sum_{i=1}^{2^n} \left(\frac{n}{1+\pi(m)}\right)^{\frac{1}{n}}$$

The famous mathematician **Pierre de Fermat** had once declared, mistakenly, that all numbers of the form $2^{2^n} + 1$ are primes, i.e., the function $f(n) = 2^{2^n} + 1$, where n is a positive integer, generates only prime numbers. However, $f(6)$, $f(8)$, $f(7)$, ... are composites. (see Fermat Numbers)

A similar method, invoving (many !) exponents was suggested by Wright in 1951, and it states that

$$g(n) = \left\lfloor 2^{2^{2^{\cdot^{\cdot^{2^{\omega}}}}}} \right\rfloor \quad \text{(a string of n exponent)}$$

where n ≥ 1, and $\omega \approx 1.9287800...$
The values of this function very grow quickly. For instance,
$$g(1) = 3,$$
$$g(2) = 13,$$
$$\text{and } g(3) = 16381.$$
However, g(4) has more than 5000 digits.

[2]

Mills' Theorem

Similarly to Wright's function, mathematician **William H. Mills** introduced a less crazy function

$$f(n) = \left\lfloor A^{3^n} \right\rfloor$$

n ≥ 1 and A is a real number.

As a matter of fact, Mills proved, in a one-page paper published in 1947, that there exists a real constant A such that $\lfloor A^{3^n} \rfloor$ is a prime number for all integers n ≥ 1. [3]

However, he did not include any clue about the value of A. His proof was nonconstructive, which means that it only showed the existence of A independently of its value.

Assuming the Riemann Hypothesis is true, the smallest possible value of A will be

$$1.3063778838630806904686144 9260\ldots$$

[4+5]

This number is called Mills' constant and is defined as the least θ for which $f_n = \lfloor \theta^{3^n} \rfloor$ is prime for all integers n. [4]
Prime numbers f_n are called Mills' Primes. If the Riemann Hypothesis is true, then

$f_1 = 2,$

$f_2 = 11,$

$f_3 = 1361,$

$f_4 = 2521008887,$

$f_5 = 16022236204009818131831320183,$

$f_6 = 4113101149215104800030529537915953170486139623539759933135949994882770404074832568499,$

...

[4+6]

More generally, it was shown that there is nothing special about the 3 in the exponent of A. Indeed, the middle exponent can be any number c ≥ 2.106, and it was shown that for any such c, there exist infinitely many A for which $\lfloor A^{c^n} \rfloor$ is prime for any natural number n.

152 Formulas for Primes

A slightly modified version of Mills' Theorem (the generalized version of it) was suggested by mathematician **Laszlo Toth** who replaced the floor function* with the ceiling function.

> *The floor function
>
> The ceiling function of a real number x, denoted ⌈x⌉, is defined as the smallest integer bigger than or equal to x.
>
> More technically, if n-1 < x ≤ n, where n is a whole number, then
>
> $$\lceil x \rceil = n.$$
>
> For example,
>
> ⌈7⌉ = 7, ⌈2.0001⌉ = 2, ⌈ −3.8⌉ = −3, ⌈π⌉ = 4, ⌈−20.20⌉= −20, ...

Toth proved that there exists a real constant B such that $\lceil B^{c^n} \rceil$ (c ≥ 2.106) is a prime number for all positive integers n.

He showed that B is approximately

$$1.24055470525201424067469515337$$

[7]

Euler's quadratic

> *O'Toole never fully comprehended what exactly was meant by the expression "quadratic prime." However, he did understand, and was fascinated by, the fact that the string 41, 43, 47, 53, 61, 71, 83, 97, ... , where each successive number was computed by increasing the difference from the previous number by 2, resulted in exactly forty consecutive prime numbers. The sequence ended only when the forty-first number in the string turned out to be a non-prime, namely 41 × 41 = 1681.*
>
> —Arthur C. Clarke and Gentry Lee, *Garden of Rama*.

In 1772, Euler discovered an interesting property of the quadratic polynomial $x^2 - x + 41$.

In fact, if we substitute the value of x with 1, 2, 3, ..., 40, then we will get the numbers

41, 43, 47, 53, 61, 71, 83, 97, 113, 131, 151, 173, 197, 223, 251, 281, 313, 347, 383, 421, 461, 503, 547, 593, 641, 691,743, 797, 853, 911, 971, 1033, 1097, 1163, 1231, 1301, 1373, 1447, 1523, 1601

which are all primes. If x = 41, then we will get 41^2 which is composite.

A few years later, Legendre noticed that a similar polynomial, $x^2 + x + 41$, is prime for x ranging from 0 to 39. However, this polynomial is called Euler's polynomial. It is worth mentioning that $x^2 + x + 41 = x(x+1) + 41$. Then, if 41 divides x or x+1, the number generated is composite and has 41 as one of its prime factors. [8]

The following is a list of the first numbers n such that $n^2 + n + 41$ is composite
40, 41, 44, 49, 56, 65, 76, 81, 82, 84, 87, 89, 91, 96, 102, 104, 109, 117, 121, 122,123,126,127,130,136,138,140, 143,147, ... [9]
This concept can be generalized into the Lucky Numbers of Euler.

A number p is said to be a Lucky Number of Euler if $n^2 - n + p$ is a prime number for integers n such that $1 \leq n \leq p-1$.

It was shown that the possible values of p are 2, 3, 5, 11, 17, or 41. There exist only 7 Lucky Numbers of Euler. [10]

Interestingly, there exist many other Prime-Generating Polynomials such as

$$n^4 - 97n^3 + 3\,294n^2 - 45\,458n + 213\,589$$

that generates prime numbers for n between 0 and 49. [11]

Another crazy prime-generating polynomial which is

$$\frac{1}{4}(n^5 - 133n^4 + 6\,729n^3 - 158\,379n^2 + 1\,720\,294n - 6\,823\,316$$

generates prime numbers for n between 0 and 56. [11]

Chapter 15

The Goldbach Conjecture

> *"It is comparatively easy to make clever guesses; indeed there are theorems, like Goldbach's Theorem, which have never been proved and which any fool could have guessed."*
>
> **Hardy, G. H. and Wright, E. M.**

Hardy, G. H. and Wright, E. M. *An Introduction to the Theory of Numbers*, 5th ed. Oxford, England: Clarendon Press, p. 19, 1979.

The Golbach Conjecture

If you look at unsolved problems in mathematics, you fill find that Goldbach's conjecture among the first on the list.

This conjecture states that:

Every even integer bigger than 4 is the sum of two prime numbers. [1]

However, the apparent simplicity of this statement hides a hard mathematical puzzle that has been challenging great number theorists for hundreds of years.

Goldbach's conjecture is named after German mathematician **Christian Goldbach** (1690–1764) born in Königsberg in Prussia (now Kaliningrad, Russia). Before joining the Imperial Academy at St. Petersburg, Russia in 1725, as a professor of mathematics and the historian of the academy, Golbdach visited many countries and met with some great mathematicians such as **Gottfried Leibniz**, ... Then, in 1728, he served as tutor to Tsar(emperor) Peter II who was crowned at the age of 11. In 1742, he joined the Russian Ministry of Foreign Affairs. [2]

Goldbach exchanged letters with many leading mathematicians of his time, especially Euler with whom he had a correspondence for more than thirty years. Indeed, it was Goldbach who encouraged Euler, in one his letters, to investigate the fifth Fermat number $F_5 = 2^{2^5} + 1$ which Euler proved to be composite by showing that it is divisible by 641. [3]

In another letter dated June 7th, 1742, Goldbach sent a letter[4] to his dear friend Euler in which he wondered about the representation of a given number as the sum of primes. He conjectured that a given number N that can be written as the sum of two prime numbers can also be written as the sum of as many primes as one wishes, until all terms are units, i.e., 1. [5]

It is worth mentioning that at the time this letter was written, 1 was considered as a prime number, so a sum of units was considered as a sum of primes.

In the same letter, Goldbach postulated that every positive integer greater than 2 is the sum of three prime numbers. The second conjecture implies the first one since if N is a positive integer bigger than two, then N = a+b+c

156 The Golbach Conjecture

where a, b, and c are prime numbers. Unless one of them is less than 2, the same process can be applied to a, b, and c,... and so forth. [4]

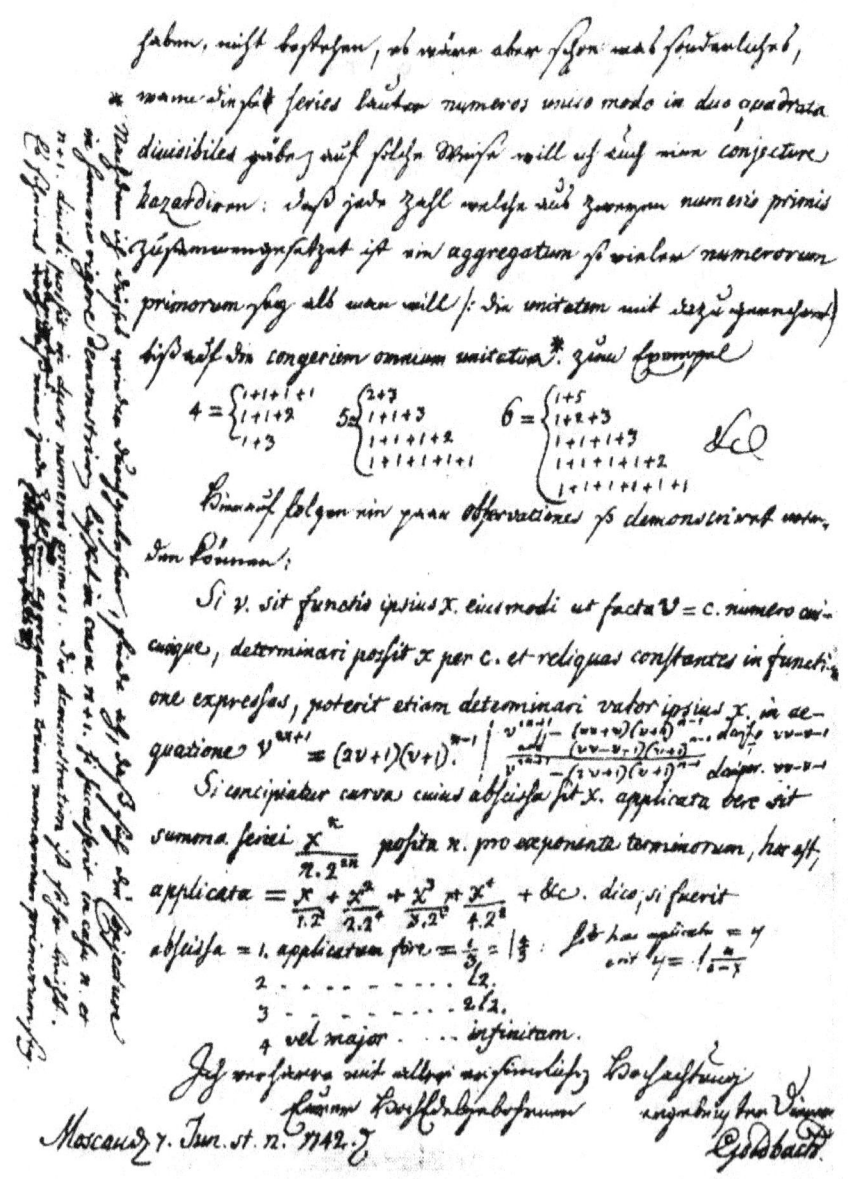

Figure 1: Letter from Goldbach to Euler, June 7th, 1742.

Euler replied to Goldbach's letter in another letter dated June 30th, 1742, in which he mentioned an old conjecture Goldbach suggested stating that every positive even integer is an aggregate of two prime numbers.

> *"Dass ... ein jeder numerus par eine summa duorum primorum sey, halte ich für ein ganz gewisses theorema, ungeachtet ich dasselbe nicht demonstriren kann."*
>
> *"That ... every even integer is a sum of two primes, I regard as a completely certain theorem, although I cannot prove it."*
>
> **Leonhard Euler**

[6]

The latter is the modern version of Goldbach's Conjecture. This modern version, stating that every even positive integer greater than 4 can be expressed as the sum of two prime numbers, is sometimes called the "strong" or "binary" Goldbach's conjecture, whereas his original postulate of the sum of three primes is called, occasionally, Goldbach's "weak" or "ternary" Conjecture, and it is also known as the 3-primes problem. Goldbach's weak Conjecture states that

Every odd number greater than 5 can be written as the sum of three primes. (A prime may be used more than once in the same sum.) [7]

If Goldbach's strong Conjecture is proven, then his weak Conjecture will be a corollary, i.e., a consequence of the first one because for example, if n − 3 can be expressed as the sum of two primes, then n is the sum of three prime numbers.

Since they were suggested in 1742, Goldbach's conjectures challenged great mathematicians and number theorists such as Euler.

At the second International Congress of Mathematicians held in Paris, France, in 1900, German mathematician **David Hilbert** (1862 –1943) gave his famous speech in which he proposed a set of the most important open problems at that time (many of them have not been completely solved to date), a list published later as Hilbert's Problems. Goldbach's Conjecture is part of the 8th problem, along with the Riemann Hypothesis. [8]

Goldbach's conjecture is also included in Landau's Problems, a list consisting of 4 unattackable problems, proposed by German mathematician **Edmund G. H. Landau** (1877 – 1938) in the 1912 fifth International Congress of Mathematicians, held in Cambridge, UK. [9]

It is important to mention that even though it is still an open problem, investigations on Goldbach's Conjecture were fruitful and have contributed to the creation and development of many powerful methods and techniques which are important not only in number theory but also in many other fields of mathematics.

In 1923, British mathematicians **Godfrey Harold Hardy** (1877 – 1947) and **John Edensor Littlewood** (1885 – 1977) showed that the ternary conjecture of Goldbach follows from the Riemann Hypothesis for all sufficiently large integers n, so if the latter is true, then so is the former. [10]

This result was improved by **Ivan Matveevich Vinogradov** (1891 – 1983) who proved the previous statement independently of the R.H. being true or false, yet, he did not give any restrictions (lower bound) on n. [11]

In 1956, $3^{14\,348\,907}$ was suggested as a lower bound to Vinogradov's result, and then, in 1989, it was replaced with $3.33 \times 10^{43\,000}$, and later, reduced to $10^{7\,194}$. [12]

In 1997, Frech mathematician **Jean-Marc Deshouillers**, American mathematician **Gove Effinger**, and Dutch computational number theorist **Herman te Riele** proved the 3-primes problem "under the Riemann Hypothesis". [13]

In 2012, there was a breakthrough in the study of Goldbach's conjectures. As a matter of fact, American mathematician **Terence Tao** proved unconditionally that every odd number greater than 1 is the sum of at most five primes. [14]

A year later, in 2013, Peruvian mathematician **Harald Andrés Helfgott** published a proof of Goldbach's weak Conjecture. [15] And thus, this 271-year old conjecture has been finally settled in the positive.

On the other hand, Goldbach's strong conjecture is still an unsolved problem. Nevertheless, many important results concerning this postulate were proven. In 1973, and then in 1978, Chinese mathematician **Chen Jingrun** published two papers in which he proved that every sufficiently large even number can be expressed as the sum of a prime number and a number that has at most two prime factors (a semiprime or a prime). This previous result is known as Chen's Theorem. Although Chen did not quantify the sufficiently large number, in 2015, Japanese mathematician **Tomohiro Yamada** showed that Chen's Theorems holds for numbers greater than

$$e^{e^{36}} \approx 1.7 \times 10^{1\,872\,344\,071\,119\,348}$$

which is an enormous lower bound.

[16]

In 1975, American mathematician **Hugh Lowell Montgomery** and British analytic number theorist **Robert Charles Vaughan** proved that "almost" all even positive integers can be expressed as the sum of two primes. (To learn more about their discovery, check their paper in the References section.)

[17]

Numerically, Portuguese computational number theorist **Tomás Oliveira e Silva** showed, in 2012-2013, that it holds for every positive integer up 4×10^{18}.

[18]

The following table summarises the upper bounds under which Goldbach's strong Conjecture was verified:

The Golbach Conjecture

Upper bound	Discoverer	
1×10^4	Honoré-Adolphe Desboves (1885)	[20.1]
1×10^5	Pipping (1938)	[20.2]
1×10^8	Stein and Stein (1965)	[20.3, 20.4]
2×10^{10}	Granville et al. (1989)	[20.5]
4×10^{11}	Sinisalo (1993)	[20.6]
1×10^{14}	Deshouillers et al. (1998)	[20.7]
4×10^{14}	Richstein (2001)	[20.8]
2×10^{16}	Oliveira e Silva (Mar. 24, 2003)	[20.9]
6×10^{16}	Oliveira e Silva (Oct. 3, 2003)	[20.10]
2×10^{17}	Oliveira e Silva (Feb. 5, 2005)	[20.11]
3×10^{17}	Oliveira e Silva (Dec. 30, 2005)	[20.12]
12×10^{17}	Oliveira e Silva (Jul. 14, 2008)	[20.13]
4×10^{18}	Oliveira e Silva (Apr. 2012)	[20.14]

[19,20]

Goldbach Partition and Goldbach Numbers

Writing a positive even number 2n as the sum of two prime numbers p and q is called the Goldbach partition of 2n, and 2n is called a Goldbach Number. Therefore, Goldbach's (strong) Conjecture can be stated as **all even integers greater than 4 are Goldbach numbers**. [21]

The following are exemples of the Goldbach partitions of some even positive integers.

4 = 2 + 2

6 = 3 + 3

8 = 3 + 5

10 = 3 + 7 = 5 + 5

...

100 = 3 + 97 = 11 + 89 = 17 + 83 = 29 + 71 = 41 + 59 = 47 + 53

...

The Golbach Conjecture 161

$1\,000 = 3 + 997 = 17 + 983 = 23 + 977 = 29 + 971 = 47 + 953 = 53 + 947$

$= 59 + 941 = 71 + 929 = 89 + 911 = 113 + 887 = 137 + 863$

$= 173 + 827 = 179 + 821 = 191 + 809 = 227 + 773 = 239 + 761$

$= 257 + 743 = 281 + 719 = 317 + 683 = 347 + 653 = 353 + 647$

$= 359 + 641 = 383 + 617 = 401 + 599 = 431 + 569 = 443 + 557$

$= 479 + 521 = 491 + 509$

The following table summarizes the number of Goldbach partitions G(n) of some even numbers n. (regardless of the order of the two primes)

n	$G(n)$	n	$G(n)$	n	$G(n)$	n	$G(n)$
4	1	20	2	36	4	100	6
6	1	22	3	38	2	200	8
8	1	24	3	40	3	392	11
10	2	26	3	42	4	1000	28
12	1	28	2	44	3	2300	49
14	2	30	3	46	4	2308	34
16	2	32	2	48	5	2310	114
18	2	34	4	50	4	2312	35

[22+23]

Clearly, G(n) increases as n gets bigger. This is obvious in the following scatterplot of G(n) for n up to 1000:

162 The Golbach Conjecture

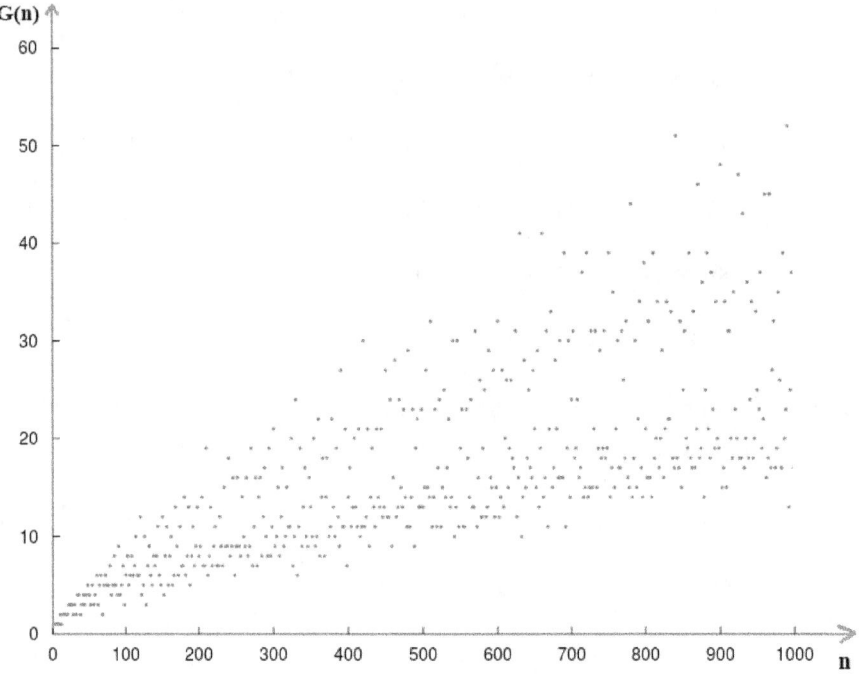

Figure 2: Scatterplot of the number of Goldbach partitions G(n) for the first 1000 integers

In their attempt to prove the Goldbach conjecture, Hardy and Littlewood gave an interesting (asymptotic) formula that estimates the number of Goldbach Partitions of a given number.

Only for the sake of enjoying the beauty of this formula that it is included here.

$$G(n) \sim 2\Pi_2 \prod_{\substack{k=2; \\ p_k | n}} \frac{p_k - 1}{p_k - 2} \int_2^n \int \frac{1}{(\ln x)^2} dx$$

where $\Pi_2 \approx 0.6601618158\ldots$ is called the twin primes constant. [24]

Related problems

A related problem to Goldbach's strong conjecture is de Polignac's Conjecture that states that every positive even number is the difference (not the sum) of two consecutive prime numbers in infinitely many ways. [25]

In 1894, French mathematician and engineer **Émile Michel Hyacinthe Lemoine** (1840 – 1912) published a slightly stronger version of Goldbach's weak Conjecture, in which he suggested that every odd number greater than or equal to 7 can be expressed as the sum of a prime p and twice a prime q, i.e., every odd number 2k+1 can be written as 2k+1 = p + 2q. [25]

For example,

$$7 = 3 + 2\times 2$$

$$9 = 3 + 2\times 3$$

$$11 = 5 + 2\times 3$$

This conjecture is called Lemoine's Conjecture, however, it is sometimes mistakenly attributed to Scottish mathematician **Hyman Levy** who was anticipated by Lemoine 69 years earlier.

In 1999, Lemoine's Conjecture was verified up to 10^9. [26]

In 2019, a blog post claimed to verify the Lemoine's Conjecture, yet it is not a reliable resource. [27]

A similar postulate to Goldbach's binary Conjecture was stated in 1984 by French mathematician **Maurice Margenstern [28]** and proved 12 years later by Italo-Swiss mathematician **Giuseppe Melfi**. [29]

This conjecture states that every even number is a sum of two practical numbers.

The Golbach Conjecture

> **Practical numbers**
>
> A number n is said to be practical if all the smaller positive integer k (k ≤ n) can be expressed as a sum of distinct proper divisors of n. [30]
>
> For example, 8 is a practical number because all integers between 1 and 8 either are divisors of 8 or can be represented as the sum of its divisors. For instance, 3 = 1 + 2, 5 = 1 + 4, 6 = 2 + 4, 7 = 1 + 2 + 4.
>
> The sequence of practical numbers goes as follows:
>
> 1, 2, 4, 6, 8, 12, 16, 18, 20, 24, 28, 30, 32, 36, 40, 42, 48, 54, 56, 60, 64, 66, 72, 78, 80, 84, 88, 90, 96, 100, ... (sequence A005153 in the OEIS)
>
> Historically, practical numbers were used in the 13th century by Italian mathematician **Leonardo Bonacci**, known as **Fibonacci**.[31] However, the name ''practical numbers'' was first attributed to this special kind of numbers in 1948 by Indian mathematician **A. K. Srinivasan**.[32]

Goldbach Conjecture in literature

Few are the mathematical problems that were the essence of successful works in literature, and Goldbach's conjecture is one of them.

In 1992, Greek author **Apostolos Doxiadis** published his famous novel *Uncle Petros and Goldbach's Conjecture,* telling a story of a young man and his uncle who tried to prove the Goldbach Conjecture. I encourage you, dear reader, to read it because additionally to the interesting plot, the author discussed many other problems in mathematics.

An anecdote about this book is that the fame it gained was partially due to the $ 1 million prize the publishers (Bloomsbury Publishing in the U.S. and Faber and Faber Limited in the UK) had offered for the first the proof of this conjecture. Although the prize could only be claimed between March 20, 2000, and March 20, 2002, and only by a resident in the US or the Uk (it went, obviously, unclaimed), it is still a significant sum of money for such an important mathematical puzzle. [33]

Goldbach's Conjecture was also involved in American writer **Isaac Asimov**'s *Sixty Million Trillion Combinations*, an interesting story about a mathematician called Vladimir Pochik whose work on the Goldbach Conjecture was stolen.

Chapter 16

Bertrand's Postulate

Chebyshev said it, but I'll say it again;

There's always a prime between n and 2n.

N. J. Fine

Schechter, B. *My Brain is Open: The Mathematical Journeys of Paul Erdős*. New York: Simon and Schuster, 1998.

Bertrand's Postulate

Prime numbers are mysterious and behave in irregular ways. To date, mathematicians have not understood thoroughly their distribution. Nevertheless, many important results have been discovered. Among those discoveries, Bertrand's postulate stands out in a crowd of major theorems concerning the study of prime numbers.

Bertrand's postulate is a theorem that states that

For any given positive integer n > 3, there is always a prime number p such that n < p < 2n – 2. [1+2+3]

From this theorem follows an important result concerning the upper bound of the gaps in the sequence of prime numbers: "**the gap to the next prime cannot be larger than the number we start our search at.**" [4]

Bertrand's postulate is sometimes stated in the following way:

There is at least a prime number between n and 2n, for any integer n > 1. [2+3]

However, this is a less restrictive, but weaker, formulation of B.P.

The following table summarises the number of primes between n and 2n-2, and between n and 2n for the first few values of n.

Bertrand's Postulate

n	The number of primes p such that n<p<2n-2. [5]	The number of primes between n and 2n (exclusive) [6]
1	0	0
2	0	1
3	0	1
4	1	2
5	1	1
6	1	2
7	1	2
8	2	2
9	2	3
10	3	4
11	3	3
12	3	4
13	3	3
14	3	3
15	3	4
16	4	5
17	4	4
18	4	4
19	3	4
20	4	4
21	4	5

Succinctly, this theorem can be expressed using the prime counting function $\pi(x)$ as

$$\pi(x) - \pi\left(\frac{x}{2}\right) \geq 1, for\ all\ x \geq 2,$$

(equivalently, $\pi(2x) - \pi(x) \geq 1, for\ all\ x \geq 2$)

where x is a positive real number (For the sake of simplicity, let x be a positive integer).

Bertrand's postulate can be differently stated as $p_{n+1} < 2p_n$, where p_n is the nth prime number.

This postulate was first suggested by French mathematician **Joseph Louis François Bertrand** (1822 –1900) who verified numerically this hypothesis for n < 3,000,000. [2]

However, it was first proved 5 years later, in 1850, by Russian mathematician **Pafnuty Lvovich Chebyshev** (1821 - 1894). **[7]** For that reason, Bertrand's postulate is sometimes called Chebyshev's Theorem.

Figure 1:
Pafnuty Lvovic
(1821 - 1894)

Figure 2:
Joseph Louis François Bertrand
(1822 - 1900)

Years later, many other mathematicians suggested different proofs than Chebyshev's.

A short but advanced proof was discovered in 1919 by the genius Indian mathematician **Srinivasa Ramanujan** (1887 – 1920). **[8]** Ramanujan not only did come out with a simple proof but also generalized Bertrand's postulate. **[9]**

Figure 3:
Srinivasa Ramanujan
(1887 - 1920)

170 Bertrand's Postulate

In his paper, Ramanujan proved the following formula:

$$\pi(x) - \pi\left(\frac{x}{2}\right) \geq n, for\ all\ x \geq R_n$$

i.e., $\pi(x) - \pi\left(\frac{x}{2}\right) \geq 1, 2, 3, 4, 5, ..., for\ all\ x \geq 2, 11, 17, 29, 41, ...$, where $\pi(x)$ is the prime counting function.

R_n are called Ramanujan Primes, and they are the smallest integers to satisfy the condition $\pi(x) - \pi\left(\frac{x}{2}\right) \geq n$, for all $x \geq R_n$. In other words, there are at least n primes between $\frac{x}{2}$ and x, whenever x is greater than or equal to R_n. [9]

The sequence of Ramanujan primes is:

2, 11, 17, 29, 41, 47, 59, 67, 71, 97, 101, 107, 127, 149, ... [10]

However, one of the most elegant and simplest proofs to this theorem was suggested by the famous Hungarian mathematician **Paul Erdős** (1913 – 1996). Indeed, his demonstration deserved to be included in the *"Proofs from The Book"**.

> *Proofs from The Book* is a famous book containing proofs of several theorems. Written in 1998 by Australian mathematician **Martin Aigner** and German mathematician **Günter M. Ziegler**, it was dedicated to the memory of the great mathematician **Paul Erdős** that devoted his entire life to mathematics, until his last hours.
> (If you don't know this man, you are missing a lot!).
>
>
>
> Figure 4:
> Paul Erdős (1913 -1996)

> During his life, Erdős often referred to The Book, in which God keeps the most perfect proof for every existing theorem. So, he described beautiful and simple proofs, especially for complicated theorems, as "proofs from The Book".
>
> *Proofs from The Book* is a less divine version of The Book, however, it is a wonderful book for people who like elegant proofs.

In his proof, Erdős used simple mathematical concepts, specifically, the binomial coefficient. Without getting into the details, he showed that the special binomial coefficient

$$\binom{2n}{n} = \frac{(2n)!}{(n!)^2}$$

called the central binomial coefficient, must have a prime factor between n and 2n; otherwise, it will be too small. He dealt with this case separately (the case where n<4000). **[4]** This proof will not be included in *Proofs from The Book*. I encourage you, dear reader, to take a look at it (you can find it easily online).

Stronger versions of Bertrand's postulate

Many mathematicians proved similar, yet stronger versions of Bertrand's postulate. For instance, in 1952, Japanese mathematician **Jitsuro Nagura** proved that for $n \geq 25$, there is always a prime number between n and $\frac{6n}{5}$. **[11]**

An improvement to this theorem was suggested by American mathematician **Lowell Schoenfeld** who, in 1976, showed that for $n \geq 2\,010\,760$, there is always a prime number between n and $\left(1 + \frac{1}{16\,579}\right)$. **[12]**

An interesting result that follows from Bertrand's postulate is the following:

172 Bertrand's Postulate

There exist a constant

$$c \approx 1.25164759777905$$

such that

$$\lfloor 2^c \rfloor, \lfloor 2^{2^c} \rfloor, \lfloor 2^{2^{2^c}} \rfloor, \ldots$$

are prime numbers. [13]

While Chebyshev showed that there exists a prime number between n and 2n, other mathematicians worked on different intervals. In fact, in 2006, mathematician **Mohamed El Bachraoui** proved that there exist at least a prime number between 2n and 3n inclusive, where n is a natural number. [14]

Previously, in 1973, Canadian mathematician **Denis Hanson** showed that there is at least a prime number between 2n and 3n. [15]

Related problems

Legendre's conjecture

We know that there exists a prime number between n and 2n, but is there any between n^2 and $(n+1)^2$?

Legendre's conjecture gives an affirmative answer. It states explicitly that

For every positive integer n there exist a prime number p between n^2 and $(n+1)^2$. [16+17]

This conjecture may seem simple. However, proving it is a tough nut to crack. For instance, it is one of Landau's problems, which are four basic, yet unattackable problems involving prime numbers. Those problems were mentioned by the German mathematician **Edmund Landau** (1877 – 1938).

[18]

It was shown that all prime and semiprime numbers p satisfy $n^2 < p < (n+1)^2$.

[17-18]

The following table summarises the number primes between n^2 and $(n+1)^2$, as well as the smallest prime between n^2 and $(n+1)^2$ for the first few values of n.

n	The smallest prime between n^2 and $(n+1)^2$ *	The number of primes between n^2 and $(n+1)^2$ **
1	2	2
2	2	5
3	2	11
4	3	17
5	2	29
6	4	37
7	3	53
8	4	67
9	3	83
10	5	101
11	4	127
12	5	149
13	5	173
14	4	197
15	6	227
16	7	257
17	5	293
18	6	331
19	6	367
20	7	401
21	7	443

* Sequence A014085 in OEIS

** Sequence A007491 in OEIS

As of August 2020, this conjecture is still unsolved.

Brocard's conjecture

A related problem to Legendre's conjecture is Brocard's conjecture, suggested by French mathematician **Pierre René J. B. H. Brocard** (1845 - 1922). This conjecture assures that

There exist at least 4 prime numbers between p_n^2 and p_{n+1}^2, where p_n is the nth prime number and n ≥ 2.

(obviously this conjecture doesn't hold for 2 and 3).

This is the same as $\pi(p_{n+1}^2) - \pi(p_n^2) \geq 4$.

[19]

The following table contains the first few numerical values of this conjecture:

n	2	3	4	5	6	7	8	9	10	11	12	13	14
Number of primes between p_n^2 and p_{n+1}^2 [20]	5	6	15	9	22	11	27	47	16	57	44	20	46

Andrica's conjecture

Andrica's conjecture is an interesting conjecture that deals with the gaps between consecutive prime numbers.

Suggested in 1986 by Romanian mathematician **Dorin Andrica**, it states that

If p_n is the nth prime number, then $\sqrt{p_{n+1}} - \sqrt{p_n} < 1$ is true for all natural numbers n.

[21+22]

If $A_n = \sqrt{p_{n+1}} - \sqrt{p_n}$, then by plotting A_n for n ≤ 100, we get:

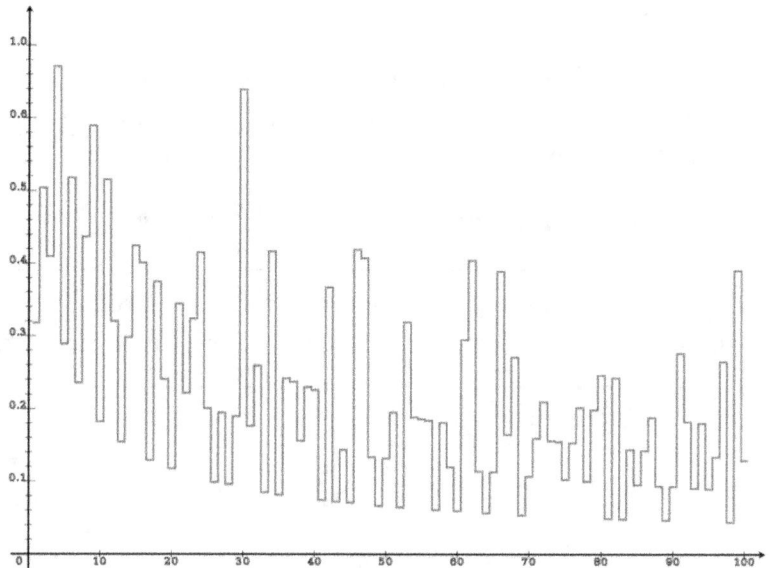

Figure 5: The graphical representation of $A_n = \sqrt{p_{n+1}} - \sqrt{p_n}$, where p_n is the nth prime number, for for the first 100 integral values of n

As you may have noticed, the highest values of this function for the first 100 primes, is $A_4 = \sqrt{11} - \sqrt{7} \approx 0.67087347$ ... This is the largest value for the first 10^5 primes.

As a piece of numerical evidence, Imran Ghory has used data on the largest prime gaps to confirm the conjecture for n up to 1.3002×10^{16}. [23]

Andrica's conjecture can be stated differently as:

$$g_n < 2\sqrt{p_n} + 1,$$

where $g_n = p_{n+1} - p_n$ (g_n is called the nth prime gap).

Note:

Sometimes, d_n is used instead of g_n, where d_n is the prime difference function defined as $d_n = p_{n+1} - p_n$.

Oppermann's conjecture

Similarly to Legendre's conjecture and Brocard's conjecture, Oppermann's conjecture, suggested in 1877 by Danish mathematician **Ludvig Henrik Ferdinand Oppermann** (1817 – 1883), states that

There exists at least a prime number between n² and n²+n.

> The product of 2 consecutive numbers is called a **pronic number** or an **oblong number**. Thus, n²+n is a pronic/oblong number.

It can be stated differently as:

$$\pi(n^2 - n) < \pi(n^2) < \pi(n^2 + n)$$

[24]

The following table contains the prime numbers existing in the gaps between $n^2 - n$ and n^2, and between n^2 and $n^2 + n$ when n ≤ 10:

n	$n^2 - n$	Primes between $n^2 - n$ and n^2	n^2	Primes between n^2 and $n^2 + n$	$n^2 + n$
2	2	3	4	5	6
3	6	7	9	11	12
4	12	13	16	17, 19	20
5	20	23	25	29	30
6	30	31	36	37, 41	42
7	42	41, 47	49	51, 53	56
8	56	59, 61	64	67, 71	72
9	72	73, 79	81	83, 89	90
10	90	93, 97	100	101, 103, 107, 109	110

[25]

If this conjecture is true, then it will imply many other conjectures. For instance, Legendre's conjecture will be proved not only for at least one prime but for at least two, between n² and (n+1)². This is because by dividing

Bertrand's Postulate

the range from n^2 to $(n+1)^2$ into two small ranges: the first range from n^2 to $n(n+1)$ and the second from $n(n+1)$ to $(n+1)^2$. Thus, if Opperman's conjecture holds, then there is at least a prime in each of the two small ranges, i.e., at least two prime numbers in the original range.

Andrica's conjecture and Brocard's conjecture will also be easily proved using Opperman's conjecture if it is true.

Chapter 17

Wilson's Theorem

Wilson's theorem is considered as an important theorem in number theory since it is strongly connected to prime numbers.

This theorem states that:

p is a prime number, if and only if $(p-1)! + 1$ is divisible by p, i.e.

$(p-1)! + 1 \equiv 0 \;(mod\; p)$

where $(p-1)! = (p-1) \times (p-2) \times (p-3) \times ... \times 3 \times 2 \times 1$. [1]

For the first few primes, we have:

n	$(n-1)! + 1$	$(n-1)! + 1 \;(mod\; n)$
2	2	0
3	3	0
4	7	3
5	25	0
6	121	1
7	721	0
8	5041	1
9	40321	1
10	362881	1
11	3628801	0
12	39916801	1
13	479001601	0
14	6227020801	1

Historically, this theorem is named after English mathematician **John Wilson** (1741 – 1793). However, Wilson was not the first to discover this wonderful fact. Arabic mathematician **Hassan Ibn-Alhaytham** (c. 965 – c. 1040) used "Wilson's Theorem" in the 10th-11th century to solve problems involving congruences.[2]

However, the first published proof of this interesting theorem was given by French mathematician **Joseph-Louis Lagrange** in 1771. Lagrange did also show the converse of this theorem, that is, he showed that if

Figure 1: Sir John Wilson (1741 - 1793)

$$(p-1)! + 1 \equiv 0 \ (mod \ p),$$

then p is a prime number. [3]

It is important to mention that before Lagrange's proof, this theorem was used only in one way, which is to say
that if p is a prime number, then

$$(p-1)! + 1 \text{ is divisible by } p.$$

Thanks to Lagrange's proof, Wilson's Theorem can be used as a primality test. It is both necessary and sufficient condition for primality. However, it is not an efficient way of proving the primeness of a given number because calculating $(n-1)!$ gets harder as n gets bigger.

Wilson's Theorem is also used in one of the prime-generating formulas.

$$f(n) = \left\lfloor \frac{n! \ (mod \ n+1)}{n} \right\rfloor (n-1) + 2$$

with n a non-zero positive integer.

180 Wilson's Theorem

It is easy to show why this function produces prime numbers. By Wilson's Theorem, if n+1 is a prime number, then $n! \equiv -1 \equiv n \pmod{n+1}$. Thus, $\left\lfloor \frac{n!(mod\ n+1)}{n} \right\rfloor = 1$, so, the function will produce n+1. Otherwise, if n+1 is not prime, then $\left\lfloor \frac{n!(mod\ n+1)}{n} \right\rfloor = 0$, and thus, the function yields 2.

However, this formula is useless at producing prime numbers for the same reason Wilson's Theorem is an inefficient primality test. [4]

Figure 2:

Hasan Ibn al-Haytham

(c. 965 – c. 1040)

Proof

A skeptical reader may be thinking of a possible proof of this theorem. There exist many. Here is an elementary one (it is a proof of only one way of the theorem, the other is left as a challenge to the reader).

The cases where n is 2 or 3 are easy to check. So, let n > 3.

If n is composite, then n and $(n-1)!$ cannot be relatively prime since gcd(n, $(n-1)!$) > 1. And this is easy to prove because if p is a factor of n such that $p < \sqrt{n}$ which means that $p < n-1$, then p must be a factor of $(n-1)!$.

However, if n is prime, then p is relatively prime with each one of the following integers: 1, 2, 3, 4, ..., n-1.

Using the notion of multiplicative inverse, for each one of the above integers i, there exists a unique positive integer j such that $i \times j \equiv 1 \pmod{n}$. By Lagrange's Theorem, $i \equiv j \pmod{n}$ if and only if $i \equiv \pm 1 \pmod{n}$, i.e., i=1 or i= n-1, since n is prime. So by excluding the 1 and n-1 from our congruence, the remaining p-3 terms on the right-hand side of the above congruence can be arranged into pairs i and j such that $i \times j \equiv 1 \pmod{n}$. Therefore, we get

$$2 \times 3 \times 4 \times 5 \times \ldots \times (n-3) \times (n-2) \equiv 1 \pmod{n}$$

By multiplying the above expression by 1 and $(n-1)$, we get

$$1 \times 2 \times 3 \times \ldots \times (n-3) \times (n-2) \times (n-1) \equiv n-1 \pmod{n}$$

$$\equiv -1 \pmod{n}$$

Equivalently, $(n-1)! \equiv -1 \pmod{n}$.

With that, our proof is finished. [5]

∎

For example, if n = 13, then

$$12! \equiv 1 \times 2 \times 3 \times 4 \times 5 \times 6 \times 7 \times 8 \times 9 \times 10 \times 11 \times 12 \pmod{13}$$

$$\equiv (1 \times 12) \times [(2 \times 7) \times (3 \times 9) \times (4 \times 10) \times (5 \times 8) \times (6 \times 11)] \pmod{13}$$

$$\equiv (-1) \times [1 \times 1 \times 1 \times 1 \times 1] \equiv -1 \pmod{13}.$$

Generalized Wilson's Theorem

The generalized version of Wilson's Theorem states that: [6]

If n is a non-zero positive integer and p is a prime number such that p ≥ n, then

$$(n-1)! \times (p-n)! \equiv (-1)^n \pmod{p}.$$

Wilson Primes

While every prime number q satisfies $(q - 1)! \equiv -1 \pmod{q}$, there exist certain primes p such that $(p - 1)! \equiv -1 \pmod{p^2}$. Those special primes p are called Wilson Primes.

We can define a Wilson Prime p differently using the Wilson Quotient of p.

The Wilson Quotient of p, denoted W_p is defined by

$$W_p = \frac{(p-1)! + 1}{p}$$

A given number p is called a Wilson Prime if $W_p \equiv 0 \pmod{p}$. [7]

It is conjectured that the number of Wilson primes is infinite. However, only 5, 13, and 563 are known [8]. Computational investigations have shown that if there exists any, then the next one has to be greater than 2×10^{11}. [9]

This limit was improved by volunteers from "mersenneforum.org" who searched for Wilson Primes up to 4×10^{11}, but nothing was found. [9+10]

It is important to mention that, heuristically, the 4th Wilson prime might be expected round about 5.10^{23}. [6]

It was proved that the number of Wilson primes in the interval [X, Y] is approximately [11]

$$\ln\left(\frac{\ln(Y)}{\ln(X)}\right)$$

Generalized Wilson Primes

Using the generalized Wilson's Theorem, Generalized Wilson Primes of order n can be defined as prime numbers p such that

$$p^2 \text{ divides } (n-1)! \times (p-n)! - (-1)^n$$

The first few Generalized Wilson primes of order n are

n	Primes p such that p^2 divides $(n-1)! \times (p-n)! - (-1)^n$ (checked up to 1,000,000)
1	5, 13, 563, ...
2	2, 3, 11, 107, 4931, ...
3	7, ...
4	10429, ...
5	5, 7, 47, ...
6	11, ...
7	17, ...
8	...
9	541, ...
10	11, 1109, ...
11	17, 2713, ...
12	...
13	13, ...
14	...

[12]

Chapter 18

The Twin Prime Conjecture

> *Mathematicians call them twin primes: pairs of prime numbers that are close to each other, almost neighbors, but between them, there is always an even number that prevents them from truly touching.*
>
> **Paolo Giordano***

* Paolo Giordano, *The Solitude of Prime Numbers: A Novel*, Penguin Books (29 March 2011)

The Twin Prime Conjecture

The twin prime conjecture is one of the most fascinating conjectures involving prime numbers.

A version of this conjecture states that **there are infinitely many consecutive primes with a difference of 2, i.e. there exist infinitely many primes p such that p+2 is also prime.** [1]

If both p and p+2 are primes, then they are called a twin prime pair. Frequently, the smallest prime in a pair of twin primes is called a Chen prime, after Chinese mathematician **Chen Jingrun**. [2]

For example, (3, 5) is a twin prime pair, and it happens to be the first*.

The next pair is (5, 7). The Chen prime of this pair, i.e., 5 also appears in (3, 5) and it is the only prime with this property.

> *Note that the pair (2, 3) is not considered as a pair of twin primes because the difference between its two elements is one, and it is the only paire of consecutive primes with this property.

The first few twin prime pairs are

(3, 5), (5, 7), (11, 13), (17, 19), (29, 31), (41, 43), (59, 61), (71, 73), (101, 103), ... [3.1]

To date, the biggest pair of twin primes, discovered on September 14, 2016, is $2996863034895 \cdot 2^{1\,290\,000} - 1$ and it has 388 342 digits. [4]

Prime numbers p such that neither p-2 nor p+2, i.e., that do not belong to a twin prime pair are called isolated primes.

The first few isolated primes are

2, 23, 37, 47, 53, 67, 79, 83, 89, 97, 113, 127, 131, 157, 163, ... [3.2]

A twin prime pair is always of the form 6k-1, 6k+1, k > 0, except for the first pair (3, 5), and this is easy to show since, upon division by 6, all integers n have remainders 0, 1, 2, 3, 4, or 5. Thus, n can be expressed as n = 6k + r,

The Twin Prime Conjecture

where r is the remainder. Obviously, 0 is eliminated, 2 and 4 are too, otherwise n would be divisible by 2. If r = 3, then n = 6k + 3 = 3(2k+1). Except when k = 0, n would be a multiple of 3 which cannot hold because we are expecting n to be prime. Therefore, we are left with 1 and 5. Either n = 6k+1 or n = 6(k+1)-1.

If you have noticed, since the prime numbers in a pair of twin primes are of the form 6k-1 and 6k+1, i.e., one less and one more than a multiple of 6, then the number between them is a multiple of 6. If you add them together, you will get a multiple of 12 (6k + 1 + 6k − 1 = 12k).

A less obvious property is that p and p+2 are twin primes if and only if

$$4[(p-1)! + 1] \equiv -p \;(mod\; p(p+2))$$

Proof

We can prove this easily using Wilson's Theorem. (See Wilson's Theorem)

Only one way of the given statement will be demonstrated. Proving the other way will be a challenge for the reader (It is not a tough challenge though because you can use Wilson's Theorem again since it is a sufficient condition for primality)

Wilson's Theorem states that p is a prime number if and only if

$$(p-1)! \equiv -1 \;(mod\; p)$$

Equivalently, $(p-1)! + 1 \equiv 0 \;(mod\; p)$. By multiplying both sides of the congruence by 4, we get,

$$4[(p-1)! + 1] \equiv 0 \;(mod\; p)$$

$$\Leftrightarrow 4[(p-1)! + 1] + p \equiv p \equiv 0 \;(mod\; p) \quad (1)$$

On the other hand, by Wilson's Theorem, if p+2 is a prime, then

$$(p+1)! + 1 \equiv 0 \;(mod\; (p+2))$$

Or, we know that

$$(p+1) \equiv -1 \;(mod\; (p+2)) \text{ and } p \equiv -2 \;(mod\; (p+2)).$$

Thus,
$$(p+1)! \equiv (p+1) \times p \times (p-1)! \equiv (-1)(-2)(p-1)$$
$$\equiv 2(p-1)! \ (mod \ (p+2))$$

Equivalently,
$$2(p-1)! \equiv (p+1)! \ (mod \ (p+2))$$

Similarly to (1),
$$4[(p-1)!+1] + p \equiv 2 \times 2(p-1)! + 4 + p \equiv 0 \ (mod \ (p+2))$$
$$\equiv 2(p+1)! + 2 + (2+p) \ (mod \ (p+2))$$
$$\equiv 2\underbrace{[(p+1)!+1]}_{\equiv 0} + \underbrace{(2+p)}_{\equiv 0} \equiv 0 \ (mod \ (p+2))$$

So,
$$4[(p-1)!+1] + p \equiv 0 \ (mod \ (p+2))$$

And we have $\quad 4[(p-1)!+1] + p \equiv 0 \ (mod \ p)$

Given that p and p+2 are coprime, we have
$$4[(p-1)!+1] + p \equiv 0 \ (mod \ p(p+2))$$

Finally,
$$4[(p-1)!+1] \equiv -p \ (mod \ p(p+2))$$

∎

A similar but a less simple formla for twin primes was discovered by Spanish mathematician **Sebastián Martín Ruiz**. He found that p and p+2 are twin prime if and only if

$$\sum_{i=1}^{p} i^a \left(\left\lfloor \frac{p+2}{i} \right\rfloor + \left\lfloor \frac{p}{i} \right\rfloor \right) = 2 + p^a + \sum_{i=1}^{p} i^a \left(\left\lfloor \frac{p+1}{i} \right\rfloor + \left\lfloor \frac{p-1}{i} \right\rfloor \right)$$

Where a ≥ 1 and ⌊ ⌋ is the floor function. [5]

188 The Twin Prime Conjecture

Historically, the name twin primes was first given to this special kind of prime numbers in 1916 by German mathematician **Paul G. S. Stäckel** (1862 –1919). [6] However, this conjecture was stated more than 50 years before as a special case of de Polignac's conjecture. In 1849, French mathematician **Alphonse de Polignac** (1826–1863) conjectured that for every positive integer n, there are infinitely many prime gaps of size 2n, i.e., there exist an infinite number of prime numbers p such that p + 2n is also prime (n is a natural number). [7]

Theorem 1 in de Polignac's paper ([7]) states that:

> "*Tout nombre pair est égal à la différence de deux nombres premiers consécutifs d'une infinité de manières...*"
>
> *Every even number is equal to the difference between two consecutive prime numbers in an infinite number of ways ...*

The case n = 1 is obviously the twin prime conjecture.

One of the first and most important results concerning this conjecture was discovered by Norwegian number theorist **Viggo Brun** (1885 – 1978) who proved in 1915 that the sum of the reciprocals of twin primes does converge, i.e., the infinite sum

$$\left(\frac{1}{3}+\frac{1}{5}\right)+\left(\frac{1}{5}+\frac{1}{7}\right)+\left(\frac{1}{11}+\frac{1}{13}\right)+\left(\frac{1}{17}+\frac{1}{19}\right)+\ldots$$

is finite and equals some real number denoted by B_2 called Brun's constant.[8] Calculating this sum using 10^{16} prime number has shown that

$$B_2 \approx 1.902160583104\ldots$$

[9]

It is important to mention that the sum of the reciprocals of prime numbers does indeed diverge to infinity, i.e., it is not finite. This was proved by Leonhard Euler in 1737. [10]

Brun's essential discovery has contributed to the development of Sieve Theory. (See Sieve of Eratosthenes) As a matter of fact, his theorem can be used to find a heuristic formula that can predict approximately the number of twin primes less than a given number n. In fact, if we let $\pi_2(n)$ to be the number of twin primes less than n (Similar to the prime counting function π), then

$$\pi_2(n) \sim K \frac{n}{[\ln(n)]^2}$$

where ~ means, without getting in many details, "approximately", and k is a definite constant that has an explicit expression. [11]

The above expression, when used with the appropriate constant k, can explain the scarcity of twin primes at larger magnitudes.

G. H. Hardy and J. E. Littlewood gave a similar but fancier formula (the following) approximating the number of twin primes less than a given number x:

$$2 \prod_{p \geq 3} \frac{p(p-2)}{(p-1)^2} \int_2^x \frac{1}{[\ln(n)]^2} dx$$

The infinite product

$$\prod_{p \geq 3} \frac{p(p-2)}{(p-1)^2}$$

is the twin prime constant and it is frequently denoted by Π_2, such that $\Pi_2 \approx 0.6601618158 \dots$ [11]

In their famous book *An Introduction to the Theory of Numbers* published in 1938, **Hardy** and **Wright** noted that the proof or disproof of the twin prime conjecture *"is at present beyond the resources of mathematics"*. [12] However, for the time being, we are so close to the proof. As a matter of fact, many major advances have been made in the study of this conjecture. In 1979, Chinese mathematician **J. R. Chen** proved that there exist an infinite number of primes p such that p+2 has at most 2 prime factors, i.e., either p+2 is prime or it is a semi-prime. [13]

Decades later, in Mai 2013, a significant breakthrough was done. American mathematician **Yitang Zhang** proved a weak version of the twin prime conjecture. Based on a previous work done by American mathematician **Daniel Alan Goldston**, Turkish mathematician **Cem Yalçın Yıldırım**, and Hungarian number theorist **János Pintz** proved that there exist infinitely many pairs of consecutive primes that differ by 7.10^7 or less. [14]

"That's only a factor of 35 million away", said Dan Goldston joking. [15]

Yet, a few months later, a significant improvement was done by British mathematician **James Maynard** who proved that there exists an infinite number of pairs of primes with a gap of at most 600. [16]

On April 14, 2014, the bound of 600 was reduced to 246 by a collaborative online project called **Polymath project 8**. Polymath project is a collaborative online platform for mathematicians to work jointly on different problems, often unsolved ones. With its two components a and b, Polymath 8 was dedicated to improving the upper bounds of small gaps between prime numbers. [17]

Nonetheless, to date, the twin prime conjecture is still an open problem.

Generalization

The twin prime conjecture is indeed a special case of many generalized conjectures that are interconnected. However, it is strongly connected to de Polignac's Conjecture.

De Polignac's Conjecture

As mentioned previously, the twin prime conjecture is a special case of de Polignac's Conjecture concerning the even gaps between consecutive prime numbers.

Cousin Primes

If the gap is 2, they are twin primes. However, if the gap is 4, they are called cousin primes.

The first few pairs of cousin primes are

(3, 7), (7, 11), (13, 17), (19, 23), (37, 41), (43, 47), (67, 71), (79, 83), (97, 101), (103, 107), (109, 113), (127, 131), (163, 167), (193, 197), (223, 227), (229, 233), (277, 281), ... [18]

Similarly to the special 5 appearing in two twin prime pairs, 7 is the only number that belongs to two different pairs of cousin primes because one of n, n+4, and n+8 will always be a multiple of 3 unless one of them is 3. In this case, we get the first two pairs of cousin primes: (3, 7) and (7, 11).

Brun's theorem can also be applied to cousin primes to prove that the sum of their reciprocals (with the initial term (3, 7) omitted) does converge and it equals a definite constant denoted by B_4. Using cousin primes up to 2^{42}, Polish physicist and mathematician **Marek Wolf** showed that

$$\left(\frac{1}{3} + \frac{1}{7}\right) + \left(\frac{1}{7} + \frac{1}{11}\right) + \left(\frac{1}{13} + \frac{1}{17}\right) + \ldots = B_4 \approx 1.1970449 \ldots$$ [19]

<u>Sexy primes</u>

If the gap between two consecutive numbers is 6, i.e., if both p and p+6 are prime, then (p, p+6) is called a pair of sexy primes. [20]

The first few pairs of sexy primes are:

(5,11), (7,13), (11,17), (13,19), (17,23), (23,29), (31,37), (37,43), (41,47), (47,53), (53,59), (61,67), (67,73), (73,79), (83,89), (97,103), (101,107), (103,109), (107,113), ... [21]

As of February 2020, the largest known pair of sexy primes of the form (p, p+6) is

p = (520461 × 2^{55931}+1) × (98569639289 × (520461 × 2^{55931}-1)^2-3)-1

p +6 = (520461 × 2^{55931}+1) × (98569639289 × (520461 × 2^{55931}-1)^2-3)+5

Those numbers 50 593-digit long. [22]

Chapter 19

The Ulam Spiral

> *Jeserac sat motionless within a whirlpool of numbers. The first thousand primes... Jeserac was no mathematician, though sometimes he liked to believe he was. All he could do was to search among the infinite array of primes for special relationships and rules which more talented men might incorporate in general laws. He could find how numbers behaved, but he could not explain why. It was his pleasure to hack his way through the arithmetical jungle, and sometimes he discovered wonders that more skillful explorers had missed. He set up the matrix of all possible integers, and started his computer stringing the primes across its surface as beads might be arranged at the intersections of a mesh.*

Arthur C. Clarke, *The City and the Stars*, 1956.

The Ulam Spiral

Looking at a sequence of prime numbers may not, at first, reveal much about those special numbers arranged in apparent chaotic order. Nevertheless, looking at prime numbers from a different perspective seems to reveal a pattern in their distribution.

In 1964, while attending a boring lecture, Polish-American mathematician **Stanisław Marcin Ulam** (1909–1984) started doodling, and his scribbles were significantly interesting. In fact, he drew a grid and placed the number 1 at its center. Starting from 1, he put the increasing natural numbers into a square spiral. Then, he crossed out the composite numbers leaving only primes. The following figure is an illustration of his doodle. [1]

Figure 1: Stanisław Ulam

194 The Ulam Spiral

100	99	98	97	96	95	94	93	92	91	90
101	64	63	62	61	60	59	58	57	56	89
102	65	36	35	34	33	32	31	30	55	88
103	66	37	16	15	14	13	12	29	54	87
104	67	38	17	4	3	2	11	28	53	86
105	68	39	18	5	0	1	10	27	52	85
106	69	40	19	6	7	8	9	26	51	84
107	70	41	20	21	22	23	24	25	50	83
108	71	42	43	44	45	46	47	48	49	82
109	72	73	74	75	76	77	78	79	80	81
110	111	112	113	114	115	116	117	118	119	120

Figure 2:
Recreation of Stanisław Ulam's scribbles

Ulam noticed something unusual and you may have noticed, too: it is the tendency of primes to fall on diagonal lines as well as horizontal and vertical fragments sometimes, or as Ulam put it: *"[it] appears to exhibit a strongly nonrandom appearance"* [2]

Investigating this strange behavior of prime numbers known for their irregularities, Ulam, with collaborators **Myron Stein** and **Mark Wells**, used one of the most powerful computers available then, the "Maniac II" at Los Alamos Scientific Laboratory, Santa Fe, New Mexico, USA, to verify numerically this observation. They started at first with 100 000 numbers and then extended their calculations to more than 10 000 000 numbers. The pattern continued in each time. Using their supercomputer, the team produced the first graphical illustration of what has been called since then the "Ulam Spiral". [2]

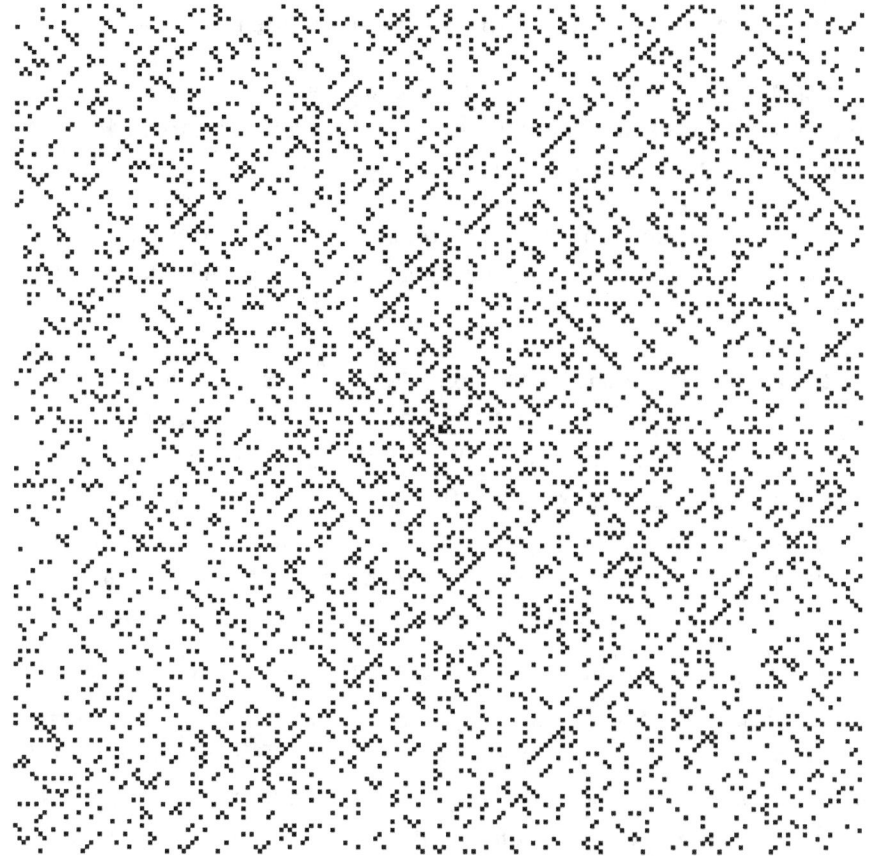

Figure 3: Ulam spiral of size 200×200

(Black dots represent prime numbers)

Evidently, prime numbers appear in a structured way. For instance, in the above illustration, diagonal, vertical, and horizontal lines are clearly visible.

What started as an observation in 1963 was popularized the following year, in March 1964 by Martin Gardner in his *Mathematical Games* column in *Scientific American*. Indeed, the Ulam Spiral appeared on the cover of that issue. [1]

In the Ulam Spiral, it is not necessary to start with 1 at the center. In fact, you can start with any natural number and you will always find the same pattern. Particularly, if you start with 41 at the center, you will get a diagonal line containing an unbroken string of 40 prime numbers. [3]

The Ulam Spiral

But how can we explain that the primes tend to fall often on diagonal lines?

In his paper, Ulam established a noted a connection between the way primes are distributed in his spiral and the prime-generating quadratics such as the famous Euler's Quadratic $n^2 - n + 41$.

More generally, prime numbers that lie on the diagonals correspond to quadratic irreducible polynomials that tend to produce primes more than others.

For example, let's consider the polynomial of $4n^2 + 8n + 5$. Writing it differently, we get $2(2n^2 + 4n + 2) + 1$. Evidently, this polynomial produces only odd numbers. Nonetheless, it will never produce a prime number for any value of n, except for 0, simply because $4n^2 + 9n + 5 = (n + 1)(4n + 5)$. However, the polynomial $4n^2 + 6n + 1$, producing also only odd numbers, is more likely to produce prime numbers. As a matter of fact, this polynomial cannot be factored into the product of two integers because the zeros of $4n^2 + 6n + 1 = 0$ are real numbers, $\frac{-3 \pm \sqrt{5}}{4}$ to be more specific. Moreover, upon division by 3 as n = 0, 1, 2, ..., the values taken by this polynomial have remainders of 1, 2, 2, 1, 2, 2,... In fact, numbers of the form $4n^2 + 6n + 1$ are not divisible by 3 for any integer value of n. (Try to prove this on your own). We continue to test how often this number is divisible by the next prime, 5. Upon division by 5 as n = 0, 1, 2,..., $4n^2 + 6n + 1$ have remainders of 1, 1, 4, 0, 4, 1, 1, 4, 0, 4, 1, 1, 4, 0, ... This affirms that the values taken by this particular polynomial are more likely to be prime numbers.

Explaining how prime numbers are connected to particular quadratic polynomials may be possible if Conjecture F is true.

The Conjecture F, suggested by English mathematicians **John Edensor Littlewood** and **Godfrey Harold Hardy** is concerned with polynomials of the form $an^2 + bn + c$, where a, b, and c are integers such that a is a non-zero positive integer. This conjecture suggests some (intuitive) conditions for which $an^2 + bn + c$ is more likely to be prime. In fact, it requires that gcd(a, b, c) = 1, the discriminant Δ ($\Delta = b^2 - 4ac$) of $an^2 + bn + c$ is not a perfect square(otherwise, the polynomial can be written as the product of two integers greater than 1), and a+b and c must not be both even. Even better, the Conjecture F suggests an asymptotic formula to estimate the number of prime numbers of the form $an^2 + bn + c$ less than a given number x, denoted P(n).

$$P(n) \sim A \frac{1}{\sqrt{a}} \frac{\sqrt{n}}{\log(n)}$$

where A is a constant defined by

$$A = \prod_p \frac{p - \omega(p)}{p - 1}$$

[4+5]

However, explaining how this formula works is a story for another time. Yet, I do encourage you, dear reader, to investigate it on your own.

Ulam Spiral has more to reveal

Instead of marking only the prime numbers in Ulam Spiral like in the previous figure, we can use dots with different sizes to represent the number of the factors of a given number in the spiral. This will reveal a hidden beautiful structure shown in the following figure.

198 The Ulam Spiral

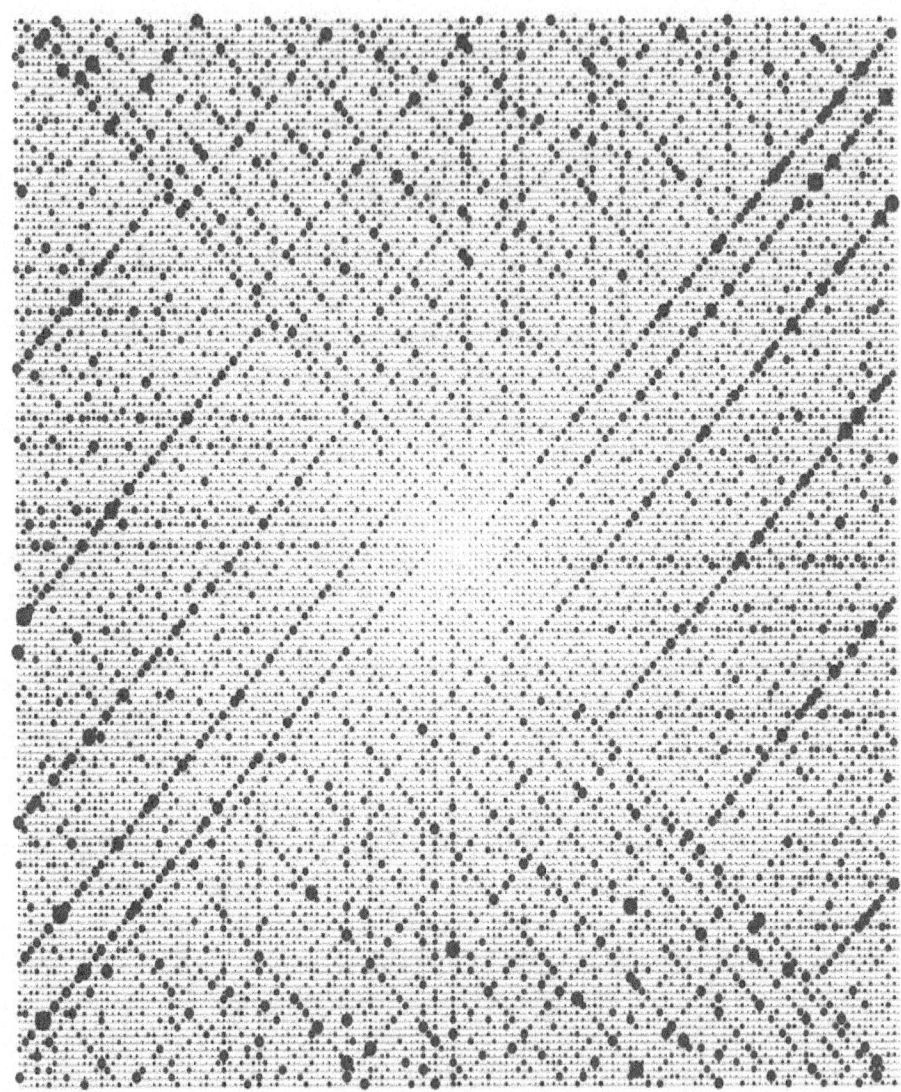

Figure 4:

150×150 Ulam spiral showing both prime and composite numbers

The size of a dot representing a number n is proportional to the number of prime factors of n.

Blue points: composite numbers
Red points: prime numbers

Variants: Klauber Triangle

32 years before Ulam's interesting doodle, American herpetologist **Laurence Monroe Klauber** (1883 – 1968) discovered a different way of illustrating the natural numbers, a way that revealed a pattern in the distribution of primes.

Instead of writing the numbers in a square spiral, **Klauber** constructed a triangle of n rows such that each row contains numbers between $(n-1)^2 + 1$ and n^2. This particular triangle is called Klauber's triangle. **[6]**

This pattern becomes clearer as the number of rows increases.

The following are illustrations of Klauber's triangle.

									1									
								2		3								
							4					7						
						8		6						13				
					9					5						21		
				14							11		12					30
			15									19		20			31	

Note: the table rendering cannot accurately represent the triangular spiral layout. Reproducing the numbers as written in the figure, by rows:

Row 1: 1
Row 2: 2 3
Row 3: 4 7
Row 4: 8 6 13
Row 5: 9 5 21
Row 6: 14 11 12
Row 7: 15 19 20 31
Row 8: 16 23 22 30 43
Row 9: 17 24 29 32 42
Row 10: 10 25 33 41 ...

Figure 5: Klauber's Triangle with 10 rows

(Orange cells contain primes, while the others are composite numbers)

The Ulam Spiral

Klauber Triangle with 200 rows

Black points represent composite numbers, the others are primes.

[7]

202 The Ulam Spiral

Similarly to the Ulam Spiral, the prime numbers in Klauber's Triangle occur often on diagonal and vertical lines. They are generated by polynomials of the form $n^2 - n + k$, where k is an integer. A particular quadratic polynomial that you may be familiar with is $n^2 + n + 41$ (Do not confuse it with Euler's quadratic polynomial $n^2 - n + 41$) is given in the following figure.

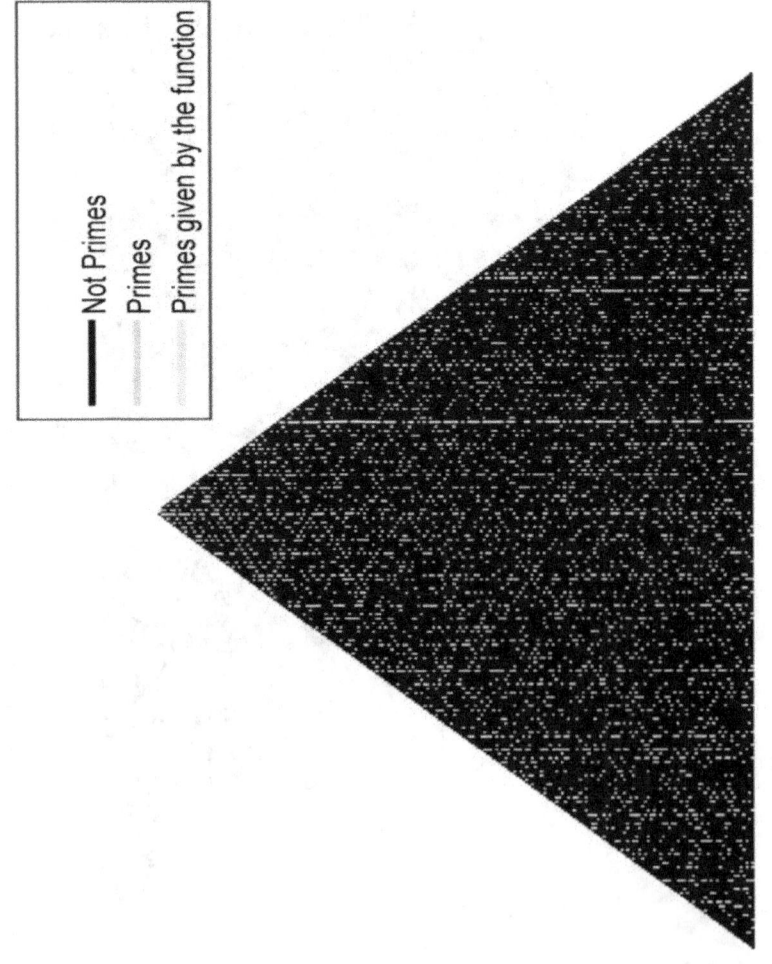

[7]

References

Chapter 1 : Primes in nature

[1] ___ John F. Dooley, *History of Cryptography and Cryptanalysis: Codes, Ciphers, and Their Algorithms*, Springer International Publishing AG (September 6, 2018)

[2] ___ Mahon, A.P. (1945). *The History of Hut Eight 1939–1945*. UK National Archives Reference HW 25/2.
Link:http://www.ellsbury.com/hut8/hut8-000.htm

[3] ___ Diffie, W.; Hellman, M.E. (November 1976). "*New directions in cryptography*". IEEE Transactions on Information Theory. **22** (6): 644–654. CiteSeerX 10.1.1.37.9720. doi:10.1109/TIT.1976.1055638.
Link: https://citeseerx.ist.psu.edu/viewdoc/summary?doi=10.1.1.37.9720

[4] ___ Rivest, Rivest, R.; Shamir, A.; Adleman, L. (February 1978). "*A Method for Obtaining Digital Signatures and Public-Key Cryptosystems*" (PDF). *Communications of the ACM*. **21** (2): 120–126. CiteSeerX 10.1.1.607.2677. doi:10.1145/359340.359342.
Link: http://people.csail.mit.edu/rivest/Rsapaper.pdf

[5] ___ Simon Singh, *The Cracking Code Book*, Harper Collins Children's Books (June 21, 2012)

[6] ___ Barker, Elaine; Dang, Quynh (2015-01-22). "NIST Special Publication 800-57 Part 3 Revision 1: Recommendation for Key Management: Application-Specific Key Management Guidance" (PDF). National Institute of Standards and Technology. p. 12. doi:10.6028/NIST.SP.800-57pt3r1

Further readings

Hellman, M. E., *The Mathematics of Public-Key Cryptography*, Scientific American, vol. 241 (August 1979), pp. 130–139.

Oded Goldreich (2001). *Foundations of Cryptography: Basic Tools*, Volume 1. Cambridge University Press. ISBN 0-521-79172-3.

Chapter 2 : Primes and Aliens

[1] ___ Wilson, John (2002). *Molecular biology of the cell: a problems approach.* New York: Garland Science. ISBN 978-0-8153-3577-1.

Link: https://archive.org/details/molecularbiolog000wils

[2] ___ Y. Fan, E. Linardopoulou, C. Friedman, E. Williams, and B. J. Trask (September 10, 2002), *Genomic Structure and Evolution of the Ancestral Chromosome Fusion Site in 2q13–2q14.1 and Paralogous Regions on Other Human Chromosome,* Cold Spring Harbor Laboratory Press.

Link: https://genome.cshlp.org/content/12/11/1651

[3] ___ Encyclopædia Britannica.

Link:https://www.britannica.com/animal/cicada

[4] ___ Cooley, J. R. 2015. *The distribution of periodical cicada (Magicicada) Brood I in 2012, with new, previously unreported populations (Hemiptera: Cicadidae).* The American Entomologist 61: 52-57.

[5] ___ Cox, R. T. & C. E. Carlton (1991). *Evidence of genetic dominance of the 13-year life cycle in periodical cicadas (Homoptera: Cicadidae: Magicicada spp.).* American Midland Naturalist. 125 (1): 63–74. doi:10.2307/2426370. JSTOR 2426370.

Link: https://www.jstor.org/stable/2426370

[6] ___ Goles, E.; Schulz, O.; Markus, M. (2001). *Prime number selection of cycles in a predator-prey model.* Complexity. 6 (4): 33–38. Bibcode:2001Cmplx...6d..33G. doi:10.1002/cplx.1040.

Link: https://onlinelibrary.wiley.com/doi/epdf/10.1002/cplx.1040

[7] ___ Teiji Sota; Satoshi Yamamoto; John R. Cooley; Kathy B. R. Hill; Chris Simon; Jin Yoshimura (2013). Independent divergence of 13- and 17-y life cycles among three periodical cicada lineages". Proceedings of the National Academy of Sciences of the United States of America. 110 (2): 6919–6924. Bibcode: 2013PNAS..110.6919S. doi:10.1073/pnas.1220060110. PMC 3637745. PMID 23509294

Link: https://www.pnas.org/content/early/2013/03/15/1220060110

Further readings

To learn more about cicadas, you can visit the following website: www.magicicada.org

Chapter 3 : Primes and cryptography

[1] ___ Courtesy NASA/JPL-Caltech.

Link : https://voyager.jpl.nasa.gov/golden-record/

[2] ___ Daniel Oberhaus, *Extraterrestrial Languages*, The MIT Press (22 octobre 2019).

[3] ___ Hans Freudenthal, *Lincos: Design of a language for cosmic intercourse*, North-Holland publishing company (January 1, 1960).

[4] ___ An interview with Dr. Yvan Dutil

Link:https://web.archive.org/web/20090704070059/http:/library.thinkquest.org/C003763/pdf/interview01.pdf

[5] ___ Carl Pomerance, *Prime Numbers and the Search for Extraterrestrial Intelligence*, Bell Labs—Lucent Technologies (2004)

[6] ___ Bill Steele, *It's the 25th anniversary of Earth's first attempt to phone E.T.* Cornell Chronicle, November 12, 1999.

Link: http://www.news.cornell.edu/releases/Nov99/Arecibo.message.ws.html

Archived version:
https://web.archive.org/web/20080802005337/http://www.news.cornell.edu/releases/Nov99/Arecibo.message.ws.html

[7] ___ Donald Goldsmith, Tobias Owen, The Search for Life in the Universe, Science Books; 3rd Edition (August 1, 2001). ISBN-13: 978-1891389160.

[8] ___ *The Arecibo Message as a 73 by 23 character message.*

Link: https://pages.uoregon.edu/jimbrau/astr123/Notes/ch28/73by23.html

Further readings

For further reading on the likelihood of the existence of extraerrestrial civilizations, and how we might detect and communicate with them:

206 References

A. G. W. Cameron, Editor, *Interstellar communication*, Benjamin, New York, 1963.

S. A. Kaplan, Editor, *Extraterrestrial civilizations; problems of interstellar communication*, (translated from the Russian edition of 1969), Israel Program for Scientific Translations, Jerusalem, 1971.

C. Sagan and F. Drake, *The search for extraterrestrial intelligence*, Scientific American, May 1975, pp. 80–89. I. S.

Shklovskii and C. Sagan, *Intelligent life in the universe*, Emerson-Adams Press, 1998. W.

Sullivan, *We are not alone: the continuing search for extraterrestrial intelligence*, Plume, 1994.

S. Webb, *If the universe is teeming with aliens ... where is everybody? Fifty solutions to the Fermi Paradox and the problem of extraterrestrial life*, Copernicus, New York, 2002.

You can read more about this topic on the Seti Institute's website: http://www.seti-inst.edu/.

Chapter 4 : How many Primes are there ?

[1] ___ T. L. Heath, *The Thirteen Books of Euclid's Elements*, vol. 2, University Press, Cambridge, 1908; 2nd ed. reprinted by Dover, New York, 1956.

[2] ___ Filip Saidak, *A NEW PROOF OF EUCLID'S THEOREM,* The American Mathematical Monthly, Vol. 113, No. 10 (Dec., 2006), pp. 937-938

Direct link: https://fermatslibrary.com/s/a-new-proof-of-euclids-theorem

Chapter 5 : The Fundamental theorem of arithmetic

[1] ___ T. L. Heath, *The Thirteen Books of Euclid's Elements*, vol. 2, University Press, Cambridge, 1908; 2nd ed. reprinted by Dover, New York, 1956.

[2] ___ Collison, M. J. (1980). *The unique factorization theorem: From Euclid to Gauss*. Mathematics Magazine 53(2):96–100.

[3] ___ Gauss, Carl Friedrich; Clarke, Arthur A. (translator into English) (1986), *Disquisitiones Arithemeticae* (Second, corrected edition), New York: Springer, ISBN 978-0-387-96254-2.

[4] ___ Ronald L. Graham, Donald E. Knuth , Oren Patashnik, *Concrete Mathematics: A Foundation for Computer Science*, (the margin of p 3), Addison Wesley (February 1994), ISBN-13: 978-0201558029.

[5] ___ Agargün, A., & Fletcher, C. (1997). *The Fundamental Theorem of Arithmetic Dissected.* The Mathematical Gazette, 81(490), 53-57. doi:10.2307/3618768

Further readings

PlanetMath's website, the Fundamental Theorem of Arithmetic,

Link: https://planetmath.org/fundamentaltheoremofarithmeticproofofthe

Weisstein, Eric W. "Fundamental Theorem of Arithmetic." From MathWorld--A Wolfram Web Resource.

Link: https://mathworld.wolfram.com/FundamentalTheoremofArithmetic.html

Chapter 6 : Sieve of Eratosthenes

[1] ___ Dictionary.com,

Link: https://www.dictionary.com/browse/sieve

[2] ___ Dictionary.com,

Link: https://www.dictionary.com/browse/sift

[3] ___ Roller, Duane W. Eratosthenes' Geography. New Jersey: Princeton University Press, 2010.

[4] ___ Samuel Horsley, *ΚΟΣ ΚΙΝΟΝ ΕΡΑΤΟΣ Θ ΕΝΟΥ Σ . or, The Sieve of Eratosthenes. Being an Account of His Method of Finding All the Prime Number*, Philosophical Transactions *(1683-1775),* Vol. 62 (1772), pp. 327-347 (21 pages). Published by the Royal Society.

Link: https://www.jstor.org/stable/106053?seq=1#metadata_info_tab_contents

[5] ___ Galaxy Magazine, p56 v18, June 1960.

Link: https://archive.org/stream/Galaxyv18n051960-06#page/n55/mode/2up/search/sieve

208 References

[6] ___ Mollin, Richard A. (2002). A *brief history of factoring and primality testing B. C. (before computers)*. Mathematics Magazine. 75 (1): 18–29. doi:10.2307/3219180. MR 2107288.

Link: https://doi.org/10.2307%2F3219180

[7] ___ Ivano Salvo, Agnese Pacifico, *Three Euler's Sieves and a Fast Prime Generator (Functional Pearl)*.

Link: https://arxiv.org/abs/1811.09840

[8] ___ A.O.L. Atkin, D.J. Bernstein, *Prime sieves using binary quadratic forms*, Math. Comp. 73 (2004), 1023-1030.

Link: https://www.ams.org/journals/mcom/2004-73-246/S0025-5718-03-01501-1/S0025-5718-03-01501-1.pdf

[9] ___

-1- D. Abdullah, R. Rahim, D Apdilah, S. Efendi, T. Tulus, and S. Suwilo, *Prime Numbers Comparison using Sieve of Eratosthenes and Sieve of Sundaram Algorithm,* Journal of Physics: Conference Series, Volume 978, 2nd International Conference on Computing and Applied Informatics 2017 28–30 November 2017, Medan, Indonesia.

Link: https://iopscience.iop.org/article/10.1088/1742-6596/978/1/012123

-2- Julian Havil, *Sundaram's Sieve*, Plus Magazine (2009).

Link: https://plus.maths.org/content/sundarams-sieve

Further readings

C. Bayes and R. Hudson, *The segmented sieve of Eratosthenes and primes in arithmetic progression*, Nordisk Tidskr. Informationsbehandling (BIT), 17:2 (1977) 121--127. MR 56:5405.

Link: https://mathscinet.ams.org/leavingmsn?url=https://doi.org/10.1007/bf01932283

Chapter 7 : The prime-counting function π

[1] ___ Weisstein, Eric W. "Prime Counting Function." From MathWorld-- A Wolfram Web Resource.
https://mathworld.wolfram.com/PrimeCountingFunction.html

References

[2] ___ Dickson, Leonard Eugene (2005). *History of the Theory of Numbers*, Vol. I: Divisibility and Primality. Dover Publications. ISBN 0-486-44232-2.

[3] ___ Caldwell, C. K. "Prime Pages. How Many Primes are There? ."

Link: https://primes.utm.edu/howmany.html

[4] ___ Online Encyclopedia for Integer Sequences.

Link: https://oeis.org/A006880

[5] ___ David Wells, *Prime Numbers: The Most Mysterious Figures in Math*, p 181, Wiley (June 2005).

[6] ___ Ingham, A. E. (1990). The *Distribution of Prime Numbers*. Cambridge University Press. pp. 2–5. ISBN 978-0-521-39789-6.

[7] ___ Hoffman, Paul (1998). The *Man Who Loved Only Numbers*. New York: Hyperion Books.. ISBN 978-0-7868-8406-3. MR 1666054.

Link: https://archive.org/details/manwholovedonlyn00hoff/page/n11/mode/2up

[8] ___ Weisstein, Eric W. "Prime Number Theorem." From MathWorld-- A Wolfram Web Resource.

Link: https://mathworld.wolfram.com/PrimeNumberTheorem.html

[9] ___ Bruce C. Berndt, *Ramanujan's Notebooks*: Part IV, Springer; 1994th Edition (December 17, 1993), ISBN-13: 978-0387941097.

[10] ___

-1- John H. Conway and R. K. Guy, *The Book of Numbers*, Copernicus, an imprint of Springer-Verlag, NY, 1995, Page 146.

-2- Online Encyclopedia for Integer Sequences

    -i- https://oeis.org/A057835

    -ii- https://oeis.org/A057752

    -iii- https://oeis.org/A006880

[11] ___

-1- Hans Riesel, *Prime Numbers and Computer Methods for Factorization*, Birkhauser Boston Inc (October 1994). ISBN-13: 978-0817637439.

210 References

-2- Kazuo AKIYAMA Tatsuhiko NAGASAWA, *On Legendre's formula and distribution of prime numbers.*

Link: https://chu-fu.ed.jp/about/pdf/issue30_pdf10.pdf

Further readings

Richard Crandall, Carl B. Pomerance, *Prime Numbers: A Computational Perspective,* Springer-Verlag New York Inc. (August 2005). ISBN-13: 978-0387252827.

Tunstrøm, Kolbjørn. (2013). *The origin of the logarithmic integral in the prime number theorem.*

Link: https://arxiv.org/abs/1311.1093

David J. Platt, Computing $\pi(x)$ analytically, Journal: Math. Comp. 84 (2015), 1521-1535.

Link: https://doi.org/10.1090/S0025-5718-2014-02884-6

Chapter 8 : Euler's Totient function

[1] ___ Long, Calvin T. (1972), *Elementary Introduction to Number Theory* (2nd ed.), p72, Lexington: D. C. Heath and Company, LCCN 77-171950

[2] ___

-1- Caldwell, C. K. "Prime Pages. Euler's phi function."

Link: https://primes.utm.edu/glossary/page.php?sort=EulersPhi

-2- Caldwell, C. K. "Prime Pages. multiplicative function."

Link: https://primes.utm.edu/glossary/page.php?sort=MultiplicativeFunction

[3] ___ Patrick Keef and David Guichard, "Introduction to Higher Mathematics".

Link: https://www.whitman.edu/mathematics/higher_math_online/section03.08.html

[4] ___ L. Euler, "*Theoremata arithmetica nova methodo demonstrata*" (An arithmetic theorem proved by a new method), *Novi commentarii academiae scientiarum imperialis Petropolitanae* (New Memoirs of the Saint-Petersburg Imperial Academy of Sciences), 8 (1763), 74–104. (The work was presented at the Saint-Petersburg Academy on October 15, 1759. A work with the same title was

presented at the Berlin Academy on June 8, 1758). Available on-line in: Ferdinand Rudio, ed., Leonhardi Euleri Commentationes Arithmeticae, volume 1, in: Leonhardi Euleri Opera Omnia, series 1, volume 2 (Leipzig, Germany, B. G. Teubner, 1915), pages 531–555. On page 531, Euler defines n as the number of integers that are smaller than N and relatively prime to N (… aequalis sit multitudini numerorum ipso N minorum, qui simul ad eum sint primi, …), which is the phi function, φ(N).

Links: -1-   https://gallica.bnf.fr/ark:/12148/bpt6k6952c/f571.image

   -2-   https://scholarlycommons.pacific.edu/euler-works/

[5] ___ Mitrinović, D. S. and Sándor, J. *Handbook of Number Theory*, p240, Dordrecht, Netherlands: Kluwer, 1995.

[6] ___ Carmichael, R. D. (1907), *On Euler's φ-function*, Bulletin of the American Mathematical Society, 13 (5): 241–243, doi:10.1090/S0002-9904-1907-01453-2.

Link: https://www.ams.org/journals/bull/1907-13-05/S0002-9904-1907-01453-2/home.html

[7] ___ Carmichael, R. D. (1922), *Note on Euler's φ-function*, Bulletin of the American Mathematical Society, 28 (3): 109–110, doi:10.1090/S0002-9904-1922-03504-5.

Link: https://www.ams.org/journals/bull/1922-28-03/S0002-9904-1922-03504-5/home.html

[8] ___ Sophia D. Merow, *Has Carmichael's Totient Conjecture Been Proven? No, No, It Has Not*, pp 759-761, Notices of the American Mathematical Society (May 2019)

Link: https://www.ams.org/journals/notices/201905/rnoti-p759.pdf

[9] ___

   -1-   Ford K, *The distribution of totients*, Ramanujan J., (2) no. 1–2: 67–151, 1998. MR1642874.

   -2-   Ford, K. *The Number of Solutions of $\phi(x) = m$*. Ann. Math. 150, 283-311, 1999.

[10] ___ Sándor, Jozsef; Crstici, Borislav (2004), *Handbook of number theory II*, Dordrecht: Kluwer Academic, pp. 228–229, ISBN 978-1-4020-2546-4, Zbl 1079.11001.

[11] ___

References

-1- Ford, K., *The Distribution of Totients*. Ramanujan J. 2, 67-151, 1998a.

-2- Ford, K., *The Distribution of Totients*, Electron. Res. Announc. Amer. Math. Soc. 4, 27-34, 1998b.

[12] ___ Weisstein, Eric W. "Sierpiński's Conjecture." From MathWorld--A Wolfram Web Resource.

Link: https://mathworld.wolfram.com/SierpinskisConjecture.html

[13] ___

-1- Weisstein, Eric W. "Totient Valence Function." From MathWorld--A Wolfram Web Resource.

Link: https://mathworld.wolfram.com/TotientValenceFunction.html

-2- Erdös, P. *Some Remarks on Euler's -Function*. Acta Math. 4, 10-19, 1958.

[14] ___ Ribenboim, Paulo (1996), *The New Book of Prime Number Records* (3rd ed.), pp 36-37, New York: Springer, ISBN 0-387-94457-5, Zbl 0856.11001.

[15] ___ Weisstein, Eric W. "Lehmer's Totient Problem." From MathWorld--A Wolfram Web Resource.

Link: https://mathworld.wolfram.com/LehmersTotientProblem.html

[16] ___

-1- Lieuwens, E. *Do There Exist Composite Numbers for Which $k * \phi(M) = M - 1$ holds?*, Nieuw Arch. Wisk. 18, 165-169, 1970.

-2- Lehmer, D. H. (1932). *On Euler's totient function*. Bulletin of the American Mathematical Society. 38: 745–751. doi:10.1090/s0002-9904-1932-05521-5. ISSN 0002-9904. Zbl 0005.34302.

Further readings

For more technical details, you can check the following:

http://gauss.math.luc.edu/greicius/Math201/Fall2012/Lectures/euler-phi.article.pdf

Florian Luca, *On the distribution of perfect totient numbers,* Journal of Integer Sequences, Vol. 9 (2006)

Link:https://web.archive.org/web/20170811183010/http://www.emis.ams.org/journals/JIS/VOL9/Luca/luca66.pdf

Rosica Dineva, *The Euler Totient, the Möbius and the Divisor Functions*, Mount Holyoke College, July 29, 2005.

Link: http://www.mtholyoke.edu/~robinson/reu/reu05/rdineva1.pdf

Paul Loomis, Michael Plytage, and John Polhill, *Summing Up the Euler φ Function*, THE MATHEMATICAL ASSOCIATION OF AMERICA.

Link: http://facstaff.bloomu.edu/jpolhill/cmj034-042.pdf

Erik R. Tou (University of Washington, Tacoma), *Math Origins: The Totient Function*, Mathematical Association of America's website.

Link: https://www.maa.org/press/periodicals/convergence/math-origins-the-totient-function

Weisstein, Eric W. "Totient Function." From MathWorld--A Wolfram Web Resource.

Link: https://mathworld.wolfram.com/TotientFunction.html

Conway, J. H. and Guy, R. K. "Euler's Totient Numbers". *The Book of Numbers*. New York: Springer-Verlag, pp. 154-156, 1996.

Caldwell, C. K. "Prime Pages. Euler's phi function."

Link: https://primes.utm.edu/glossary/page.php?sort=EulersPhi

Weisstein, Eric W. "Carmichael's Totient Function Conjecture." From MathWorld--A Wolfram Web Resource.

Link: https://mathworld.wolfram.com/CarmichaelsTotientFunctionConjecture.html

Proceeding of the American Mathematical Society, Volume 43, Number2, April 1974, On Carmichael's Conjecture, Carl Pomerance.

Link: https://math.dartmouth.edu/~carlp/PDF/carmichaelconjecture.pdf

Chapter 9 : Mersenne Primes

[1] ___ Utriusque Arithmetices epitome, Regius, Hudalricus (1536).

References

Link of the book on Google Books:
https://books.google.tn/books?id=hs85AAAAcAAJ&dq=Utriusque%20Arithmetices%20epitome&hl=fr&pg=PP5#v=onepage&q&f=false

[2] ___ Caldwell, C. K. "Prime Pages. Mersenne Primes."

Link: https://primes.utm.edu/mersenne/index.html

[3] ___ O'Connor, John J.; Robertson, Edmund F., *Pietro Antonio Cataldi*, MacTutor History of Mathematics archive, University of St Andrews.

Link: https://mathshistory.st-andrews.ac.uk/Biographies/Cataldi/

[4] ___ L. E. Dickson, *History of the theory of numbers* (Vol. 1, p28), Carnegie Institute of Washington, 1919. Reprinted by Chelsea Publishing, New York, 1971.

[5] ___ Caldwell, C. K. "Prime Pages. Mersenne Primes."

Link: https://primes.utm.edu/mersenne/index.html

[6] ___ Dora Musielak, *Prime Mystery: The Life and Mathematics of Sophie Germain: Revolutionary Mathematician*, (second edition) Springer Nature Switzerland AG.

[7] ___ Agoh, Takashi, *On Sophie Germain primes*, Tatra Mt. Math. Publ., 20 (2000) 65--73. Number theory (Liptovský Ján, 1999).

[8] ___ http://www.primegrid.com/forum_thread.php?id=6686

[9] ___
-1- D. H. Lehmer, *On certain chains of primes*, Proc. Lond. Math. Soc., series 3, 14a (1965) 183--186.

-2- Caldwell, C. K. "Prime Pages. Cunningham chain."
https://primes.utm.edu/glossary/page.php?sort=CunninghamChain

[10] ___ Caldwell, C. K. "Prime Pages. Euler and Lagrange on Mersenne Divisors."

https://primes.utm.edu/notes/proofs/MerDiv2.html

[11] ___ GIMPS Project, https://www.mersenne.org/various/history.php

[12] ___ GIMPS Project, https://www.mersenne.org/primes/

[13] ___ The Electronic Frontier Foundation (EFF)
https://www.eff.org/press/releases/big-prime-nets-big-prize

References 215

[14] ___

-1- "mt_rand — Generate a better random value". PHP Manual. Retrieved March 16th, 2020.

Link: https://www.php.net/manual/en/function.mt-rand.php

-2- "random — Generate pseudo-random numbers — Python 3.8.3 documentation". Python 3.8.3 documentation. Retrieved 2020-06-23.

Link: https://docs.python.org/3/library/random.html

-3- "Random Number Generators". CRAN Task View: Probability Distributions.

Link: https://cran.r-project.org/web/views/Distributions.html

-4- " "Random" class documentation". Ruby 1.9.3 documentation. Retrieved 2012-05-29.

Link: https://ruby-doc.org/core-1.9.3/Random.html

[15] ___ John Savard. *The Mersenne Twister*. A subsequent paper, published in the year 2000, gave five additional forms of the Mersenne Twister with period $2^{19937}-1$. All five were designed to be implemented with 64-bit arithmetic instead of 32-bit arithmetic.

Link: http://www.quadibloc.com/crypto/co4814.htm

[16] ___ Encyclopædia Britannica,

https://www.britannica.com/topic/Tower-of-Hanoi

[17]___ Petković, Miodrag (2009*). Famous Puzzles of Great Mathematicians*, AMS Bookstore. p. 197. ISBN 978-0-8218-4814-2.

[18]___

-1- Gerstein, Larry (2012), *Introduction to Mathematical Structures and Proofs*, Undergraduate Texts in Mathematics, Springer, Theorem 6.94, p. 339, ISBN 9781461442653.

Link of the book on Google Books:
https://books.google.tn/books?id=qK9y768b1NQC&lpg=PA339&hl=fr&pg=PA339#v=onepage&q&f=false

-2- Caldwell, Chris K., "*A proof that all even perfect numbers are a power of two times a Mersenne prime*", Prime Pages.

Link:
https://primes.utm.edu/notes/proofs/EvenPerfect.htmlhttps://primes.utm.edu/notes/proofs/EvenPerfect.html

[19] ___ R. P. Brent, G. L. Cohen and H. J. J. te Riele, *Improved techniques for lower bounds for odd perfect numbers*, Math. Comp., 57:196 (1991) 857--868.

[20] ___ L. E. Dickson, *History of the theory of numbers*, Carnegie Institute of Washington, 1919. Reprinted by Chelsea Publishing, New York, 1971.

[21] ___ Bajnok, Béla (2013), *An Invitation to Abstract Mathematics*, Undergraduate Texts in Mathematics, Springer, ISBN 978-1-4614-6636-9.

Link of the book on Google Books:
https://books.google.tn/books?id=qK9y768b1NQC&lpg=PA339&hl=fr&pg=PA339#v=onepage&q&f=false

[22] ___ M. Brooke, *On the digital roots of perfect numbers*, Math. Mag., 34:2 (1960) 100.

[23] ___ L. E. Dickson, *History of the theory of numbers*, Carnegie Institute of Washington, 1919. Reprinted by Chelsea Publishing, New York, 1971.

Chapter 10 : Pierre de Fermat

[1] ___

-1- https://www.famousscientists.org/

-2- https://www.britannica.com/biography/Pierre-de-Fermat

-3- *The Hutchinson Dictionary of Scientific Biography, Pierre de Fermat (1601-1665)*, Massachusetts, 1988.

[2] ___ Pellegrino, Dana. Pierre de Fermat, (2000)

Link: https://sites.math.rutgers.edu/~cherlin/History/Papers2000/pellegrino.html

[3] ___ O'Connor, John J.; Robertson, Edmund F, *Pierre de Fermat*, MacTutor History of Mathematics archive, University of St Andrews.

Link: https://mathshistory.st-andrews.ac.uk/Biographies/Fermat

[4] ___
-1- http://www.livingreviews.org/lrr-2004-9

-2- Volker Perlick, *Gravitational Lensing from a Spacetime Perspective*, *Living Rev. Relativity* **7**, (2004), 9.
Links:
https://web.archive.org/web/20160303235551/http://relativity.livingreviews.org/open?pubNo=lrr-2004-9&page=articlesu9.html

[5] ___

-1- Dunham, William. *Number theory – Number theory in the East*. Encyclopedia Britannica.

Link: https://www.britannica.com/science/number-theory

-2- O'Connor, J. J.; Robertson, E. F. (February 2002). *Pell's Equation*. School of Mathematics and Statistics, University of St Andrews, Scotland. Retrieved 13 July 2020.

Link: https://mathshistory.st-andrews.ac.uk/HistTopics/Pell/#:~:text=where%20n%20is%20a%20given,related%20to%20Pell's%20equation.

[6] ___ Fermat, Pierre (1894), Tannery, P.; Henry, C. (eds.), *Oeuvres de Fermat. Tome 2: Correspondance*, Paris: Gauthier-Villars, p 514 (in French)

Source: Bibliothèque nationale de France, département Sciences et techniques, 2011-314981.

Link: https://gallica.bnf.fr/ark:/12148/bpt6k6213616t/f280.item.texteImage.zoom

[7] ___ Edwards, Harold M. (2000), "1.6 Fermat's one proof", *Fermat's Last Theorem: A Genetic Introduction to Algebraic Number Theory, Graduate Texts in Mathematics*, Springer, pp. 10–14, ISBN 978-0-387-95002-0

Link on Google Books:

https://books.google.tn/books?id=_IxN-5PW8asC&pg=PA10&redir_esc=y#v=onepage&q&f=false

[8] ___ Kato, Kazuya, Saitō, Takeshi (2000), *Number Theory: Fermat's dream, Translations of mathematical monographs*, translated by Nobushige Kurokawa, American Mathematical Society, p. 17, ISBN 978-0-8218-0863-4

[9] ___ Weisstein, Eric W. "*Fermat's Polygonal Number Theorem*". MathWorld.

Link: https://mathworld.wolfram.com/FermatsPolygonalNumberTheorem.html

[10] ___ Encyclopædia Britannica.

218   References

Link: https://www.britannica.com/topic/number-game/Paradoxes-and-fallacies#ref396120

[11] ___ Daniela Betancourt, Timothy Park, *Polygonal numbers*, (Project for MA 341 Introduction to Number Theory, Instractor: Kalin Kostadinov) Boston University, Summer Term 2009.

Link: http://math.bu.edu/people/kost/teaching/MA341/PolyNums.pdf

[12] ___

-1- Heath, Thomas Little. *A History of Greek Mathematics: From Thales to Euclid*. Courier Dover Publication 1981.

-2- Heath, Sir Thomas Little (1910), *Diophantus of Alexandria; a study in the history of Greek algebra*, Cambridge University Press, p. 188.

Link: https://archive.org/details/diophantusofalex00heatiala

[13] ___ Weisstein, Eric W. "Pollock's Conjecture." From MathWorld--A Wolfram Web Resource. https://mathworld.wolfram.com/PollocksConjecture.html

[14] ___ Dickson, L. E. (June 7, 2005). *History of the Theory of Numbers*, Vol. II: Diophantine Analysis. Dover. pp. 22–23. ISBN 0-486-44233-0.

[15] ___ AoPS Online (Art of Problem Solving)

Link: https://artofproblemsolving.com/wiki/index.php/Fermat

[16] ___ L. E. Dickson, *History of the Theory of Numbers*, Vol. II, Ch. VI, p. 227.

[17] ___ Jahnavi Bhaskar, *Sum of two squares*, (2008)

Link: https://www.math.uchicago.edu/~may/VIGRE/VIGRE2008/REUPapers/Bhaskar.pdf

[18] ___ Dickson, Leonard Eugene (1920). "Ch. VI: Sum of two squares". *History of the Theory of Numbers*, Vol. II. pp. 227–228.

Link: https://archive.org/stream/historyoftheoryo02dickuoft#page/226/mode/2up

[19] ___ Colin R. Fletcher, *A reconstruction of the Frenicle-Fermat correspondence of 1640,* Historia Mathematica, Volume 18, Issue 4, November 1991, Pages 344-351.

[20] ___ Euler's two papers

References   219

-1- De numerus qui sunt aggregata quorum quadratorum. (Novi commentarii academiae scientiarum Petropolitanae 4 (1752/3), 1758, 3-40)

-2- Demonstratio theorematis FERMATIANI omnem numerum primum formae 4n+1 esse summam duorum quadratorum. (Novi commentarii academiae scientiarum Petropolitanae 5 (1754/5), 1760, 3-13)

[21] ___ Weisstein, Eric W. "Taniyama-Shimura Conjecture." From MathWorld--A Wolfram Web Resource.

Link: https://mathworld.wolfram.com/Taniyama-ShimuraConjecture.html

[22] ___ The Abel Prize 2016.

Link:https://www.abelprize.no/c67107/binfil/download.php?tid=67059

[23] ___ Yu. I. Manin, Alexei A., *Introduction to Modern Number Theory: Fundamental Problems, Ideas and Theories,* Springer Science & Business Media, (March 2006).

Link on Google Books:
https://books.google.tn/books?id=wvK586IxaxwC&pg=PA341&dq=%22Cubum+au tem+in+duos+cubos,+aut+quadratoquadratum%22&hl=en&ei=Jig_Tc2hCIH_8Aaf2 LivCg&sa=X&oi=book_result&ct=result&sqi=2&redir_esc=y#v=onepage&q=%22 Cubum%20autem%20in%20duos%20cubos%2C%20aut%20quadratoquadratum%2 2&f=false

[24] ___ O'Connor, John J.; Robertson, Edmund F, *Fermat's Last Theorem,* MacTutor History of Mathematics archive, University of St Andrews.

Link: https://mathshistory.st-andrews.ac.uk/HistTopics/Fermat

[25] ___ Laubenbacher R, Pengelley D (2007). *Voici ce que j'ai trouvé: Sophie Germain's grand plan to prove Fermat's Last Theorem,* (PDF). January 24, 2010.

Link: https://web.archive.org/web/20130405163013/http:/www.math.nmsu.edu/~davidp/g ermain.pdf

[26] ___ David Wells, "Fermat, Pierre de (1607–1665) ", *Prime Numbers: The Most Mysterious Figures in Math,* p 99, Wiley (June 2005).

[27] ___ Pomerance, Carl (2008), *Computational Number Theory*, in Gowers, Timothy; Barrow-Green, June; Leader, Imre (eds.), The Princeton Companion to Mathematics, Princeton University Press, pp. 361–362, ISBN 978-0-691-11880-2.

[28] ___ Online Encyclopedia for Integer Sequences.

220 References

Link: https://oeis.org/A214618

[29] ___ R. K. Guy, *Unsolved problems in number theory*, Springer-Verlag, New York, NY, 1994. ISBN 0-387-94289-0.

[30] ___ P. Mihailescu, *A class number free criterion for Catalan's conjecture*, J. Number Theory, 99:2 (2003) 225--231.

Link: https://doi.org/10.1016/S0022-314X(02)00101-4

[31] ___ Caldwell, C. K. "Prime Pages. Catalan's problem."

Link: https://primes.utm.edu/glossary/page.php?sort=CatalansProblem

[32] ___ Beal Conjecture's website.

Link: https://www.bealconjecture.com/

[33] ___ Beal Prize, American Mathematical Society.

Link: https://www.ams.org/prizes-awards/paview.cgi?parent_id=41

[34] ___ Fermat, Pierre (1894), Tannery, P.; Henry, C. (eds.), *Oeuvres de Fermat. Tome 2: Correspondance*, Paris: Gauthier-Villars, pp. 206–212 (in French).

Link: https://archive.org/stream/oeuvresdefermat02ferm#page/210/mode/2up

[35] ___ Mahoney, Michael Sean (1994), *The Mathematical Career of Pierre de Fermat, 1601–1665* (2nd ed.), Princeton University Press, ISBN 978-0-691-03666-3 (English Version p295).

[36] ___ Long, Calvin T. (1972), *Elementary Introduction to Number Theory* (2nd ed.), Lexington: D. C. Heath and Company, LCCN 77171950.

[37] ___ Riesel, Hans (1994). *Prime Numbers and Computer Methods for Factorization*. Progress in Mathematics. 126 (second ed.). Boston, MA: Birkhäuser. ISBN 978-0-8176-3743-9. Zbl 0821.11001.

[38] ___

-1- Weisstein, Eric W. "*Fermat's Little Theorem Converse.*" From MathWorld--A Wolfram Web Resource.

Link: https://mathworld.wolfram.com/FermatsLittleTheoremConverse.html

-2- Wagon, S. Mathematica in Action. New York: W. H. Freeman, pp. 278-279, 1991.

-3- P. S. Bruckman, *A converse of Fermat's Little Theorem*, International Journal of Mathematical Education in Science and Technology, Volume 38, 2007 - Issue 4, pp 554-555.

Link: https://doi.org/10.1080/00207390701228419

[39] ___ Burton, David M. (2011), *The History of Mathematics, An Introduction* (7th ed.), p 514, McGraw-Hill, ISBN 978-0-07-338315-6.

[40] ___ Ore, Oystein (1988), *Number Theory and Its History*, p273, Dover, ISBN 978-0-486-65620-5.

Link: https://archive.org/details/numbertheoryitsh0000orey

[41] ___ Leonhard Euler, *Theorematum quorundam ad numeros primos spectantium demonstratio (A proof of certain theorems regarding prime numbers)*, Commentarii academiae scientiarum Petropolitanae, 8: 141–146. (published in 1741)

For further details on this paper, including an English translation, see **The Euler Archive.**

Link: https://scholarlycommons.pacific.edu/euler-works/54/

[42] ___ Hardy, G. H.; Wright, E. M. (2008), "Fermat's Theorem and its Consequences", *An Introduction to the Theory of Numbers* (6th ed.), Oxford University Press, ISBN 978-0-19-921986-5.

[43] ___ Ivory, James (1806), *Demonstration of a theorem respecting prime numbers*, New Series of the Mathematical Depository, 1 (II): 6–8.

[44] ___ Weisstein, Eric W. "Fermat Quotient." From MathWorld--A Wolfram Web Resource.

Link: https://mathworld.wolfram.com/FermatQuotient.html

[45] ___ Paulo Ribenboim, *13 Lectures on Fermat's Last Theorem* (1979), especially pp. 152, 159-161.

[46] ___ Gotthold Eisenstein, *Neue Gattung zahlentheoret. Funktionen, die v. 2 Elementen abhangen und durch gewisse lineare Funktional-Gleichungen definirt werden, Bericht über die zur Bekanntmachung geeigneten Verhandlungen der Königl.* Preuß. Akademie der Wissenschaften zu Berlin 1850, 36-42.

[47] ___

-1- Caldwell, C. K. "Prime Pages. Wieferich prime."
https://primes.utm.edu/glossary/page.php?sort=WieferichPrime

References

-2- Weisstein, Eric W. "Wieferich Prime." From MathWorld--A Wolfram Web Resource.

Link: https://mathworld.wolfram.com/WieferichPrime.html

[48] ___ Wieferich, A. (1909), *Zum letzten Fermat'schen Theorem*, Journal für die reine und angewandte Mathematik (in German), 1909 (136): 293–302, doi:10.1515/crll.1909.136.293.

Link: https://doi.org/10.1515%2Fcrll.1909.136.293

[49] ___ Weisstein, Eric W. "Mirimanoff's Congruence." From MathWorld--A Wolfram Web Resource.

Link: https://mathworld.wolfram.com/MirimanoffsCongruence.html

[50] ___ Online Encyclopedia for Integer Sequences.

Link: https://oeis.org/A001220

[51] ___ Richard Fischer, fermatquotient.com.

Link: http://www.fermatquotient.com/FermatQuotienten/FermQSort.txt

[52] ___ Crandall, Richard E.; Dilcher, Karl; Pomerance, Carl (1997), *A search for Wieferich and Wilson primes* (PDF), Mathematics of Computation, 66 (217): 433–449, Bibcode:1997MaCom..66..433C, doi:10.1090/S0025-5718-97-00791-6.

[53] ___ Lehmer, D. H. *On Fermat's Quotient, Base Two*. Math. Comput. 36, 289-290, 1981.

[54] ___ Wilfrid Keller; Jörg Richstein (2005), *Solutions of the congruence $a^{p-1} \equiv 1 \pmod{p^r}$*, (PDF), Mathematics of Computation, 74 (250): 927–936, doi:10.1090/S0025-5718-04-01666-7.

Link: https://www.ams.org/journals/mcom/2005-74-250/S0025-5718-04-01666-7/S0025-5718-04-01666-7.pdf

[55] ___

-1- Online Encyclopedia for Integer Sequences.

Link: https://oeis.org/A282293

-2- Weisstein, Eric W. "Double Wieferich Prime Pair." From MathWorld--A Wolfram Web Resource.

Link: https://mathworld.wolfram.com/DoubleWieferichPrimePair.html

[56] ___ Online Encyclopedia for Integer Sequences.

Link: https://oeis.org/A077816

[57] ___ Agoh, T.; Dilcher, K.; Skula, L. (1997), *Fermat Quotients for Composite Moduli*, Journal of Number Theory, 66 (1): 29–50, doi:10.1006/jnth.1997.2162

Link: https://doi.org/10.1006%2Fjnth.1997.2162

[58] ___

-1- Michal Krizek, Florian Luca, Lawrence Somer, *17 Lectures on Fermat Numbers: From Number Theory to Geometry*, Springer Science & Business Media, 2013.

-2- Online Encyclopedia for Integer Sequences

Link: https://oeis.org/A000215

[59] ___

-1- Letter dated August[?], 1640to Fenicle de Bessy (Oeuvres*, p206);

-2- Letter dated 25 December 1640 to Mersenne (Oeuvres*, pp 212-213).

* Fermat, Pierre (1894), Tannery, P.; Henry, C. (eds.), *Oeuvres de Fermat. Tome 2: Correspondance*, Paris: Gauthier-Villars (in French).

[60] ___ Caldwell, C. K. "Prime Pages. Fermat Number."

Link: https://primes.utm.edu/glossary/page.php?sort=FermatNumber

[61] ___ Křížek M., Luca F., Somer L. (2002) "The Most Beautiful Theorems on Fermat Numbers", *17 Lectures on Fermat Numbers*. CMS Books in Mathematics. Springer, New York, NY.

Link: https://doi.org/10.1007/978-0-387-21850-2_4

[62] ___ Caldwell, C. K. "Prime Pages. Goldbach's Proof of the Infinitude of Primes (1730)."

Link: https://primes.utm.edu/notes/proofs/infinite/goldbach.html

[63] ___ PrimeSearch: Distributed Search for Fermat Prime Number Divisors.

Link: http://www.fermatsearch.org/

References

[64] ___ Wilfrid Keller, ProthSearch.

Link: http://www.prothsearch.com/fermat.html#Summary

[65] ___ Boklan, Kent D.; Conway, John H. (2016). *Expect at most one billionth of a new Fermat Prime!*. arXiv:1605.01371.

Link: https://arxiv.org/abs/1605.01371

[66] ___ Online Encyclopedia for Integer Sequences

Link: https://oeis.org/A050922

[67] ___ Weisstein, Eric W. "Fermat Number." From MathWorld--A Wolfram Web Resource.

Link: https://mathworld.wolfram.com/FermatNumber.html

[68] ___ Wantzel, M. L. *Recherches sur les moyens de reconnaître si un problème de géométrie peut se résoudre avec la règle et le compas*, J. Math. pures appliq. 1, 366-372, 1836.

[69] ___ Weisstein, Eric W. "Generalized Fermat Number." From MathWorld--A Wolfram Web Resource.

Link: https://mathworld.wolfram.com/GeneralizedFermatNumber.html

[70] ___ Caldwell, C. K. "Prime Pages. Generalized Fermat."

Link: https://primes.utm.edu/top20/page.php?id=12

Further readings

Howard Eves, *An Introduction to the History of Mathematics*, Saunders College Publishing, Fort Worth, TX, 1990.

Adleman, L.M., Heath-Brown, D.R. *The first case of Fermat's last theorem.* Invent Math 79, 409–416 (1985).

Link: https://doi.org/10.1007/BF01388981

Joseph H. Silverman, *A Friendly Introduction to Number Theory*, Koll Publishers, New Jersey, 1997.

An excellent book to learn about Figurate Numbers

Elena Deza, M. Deza, *Figurate Numbers,* World Scientific (2012)

Link on Google Books:

https://books.google.tn/books?id=cDxYdstLPz4C&pg=PA1&hl=fr&source=gbs_toc_r&cad=4#v=onepage&q&f=false

Fermat's Last theorem

Simon Singh, *Fermat's Last Theorem*, HarperPress (November 22, 2012).

P. Ribenboim, *Fermat's last theorem for amateurs*, Springer-Verlag, 1999. New York, NY, pp. xiv+407, ISBN 0-387-98508-5. MR 2001h:11036

Link: http://www.ams.org/mathscinet-getitem?mr=2001h:11036

Gerd Faltings, *Proof of Fermat's Last Theorem by R.Taylor and A.Wiles*, Notices of the AMS, volume 42, number 7, July 1995.

Link: http://www.ams.org/notices/199507/faltings.pdf

Chapter 11 : Riemann Hypothesis

[1] ___ Bombieri, Enrico. *The Riemann Hypothesis – official problem description*, (PDF), Clay Mathematics Institute.

(This paper also contained the Riemann hypothesis, a conjecture about the distribution of complex zeros of the Riemann zeta function that is considered by many mathematicians to be the most important unsolved problem in pure mathematics.)

Link: http://www.claymath.org/sites/default/files/officialproblemdescription.pdf

[2] ___ Ingham, A.E. (1932), *The Distribution of Prime Numbers*, Cambridge Tracts in Mathematics and Mathematical Physics, 30, Cambridge University Press.

[3] ___

-1- The First 10,000 Primes.

Link: https://primes.utm.edu/lists/small/10000.txt

-2- Created from the first list using a simple python program

[4] ___ Eric Barkan, David Sklar, *On Riemanns Nachlass for Analytic Number Theory: A translation of Siegel's Uber*.

Link: https://arxiv.org/abs/1810.05198

[5] ___ Encyclopædia Britannica.

References

Link: https://www.britannica.com/biography/Bernhard-Riemann

[6] ___ Dan Rockmore, *Stalking the Riemann Hypothesis: The Quest to Find the Hidden Law of Prime Numbers*, p 66, ed. Vintage (Mai 9, 2006), ISBN-13: 978-0375727726.

[7] ___ O'Connor, John J.; Robertson, Edmund F., *Bernhard Riemann*, MacTutor History of Mathematics archive, University of St Andrews.

Link: https://mathshistory.st-andrews.ac.uk/Biographies/Riemann

[8] ___ Encyclopædia Britannica.

Link: https://www.britannica.com/science/analysis-mathematics/Complex-analysis

[9] ___ Weisstein, Eric W. "Complex Analysis." From MathWorld--A Wolfram Web Resource.

Link: https://mathworld.wolfram.com/ComplexAnalysis.html

[10] ___ Hahn, Harry K. *The Ordered Distribution of Natural Numbers on the Square Root Spiral*. arXiv:0712.2184

Link: https://arxiv.org/abs/0712.2184

[11] ___ Tobias Dantzig, *Number, the language of science; A critical survey written for the cultured non-mathematician*, pp.187-213, New York, The Macmillan Company, 1930.

[12] ___ Titu Andreescu, Dorin Andrica, *Complex Numbers from A to ... Z*, Birkhäuser; 2nd Edition (February 17, 2014).

[13] ___ Weisstein, Eric W. "Imaginary Number." From MathWorld--A Wolfram Web Resource.

Link: https://mathworld.wolfram.com/ImaginaryNumber.html

[14] ___ Giaquinta, Mariano; Modica, Giuseppe, *Mathematical Analysis: Approximation and Discrete Processes* (illustrated ed.). Springer Science & Business Media(2004). p. 121. ISBN 978-0-8176-4337-9.

Link on Google Books (Extract of page 121):
https://books.google.tn/books?id=Z6q4EDRMC2UC&pg=PA121&redir_esc=y#v=onepage&q&f=false

[15] ___ Encyclopædia Britannica.

https://www.britannica.com/science/Riemann

[16] ___ Jørgen Veisdal, *The Riemann Hypothesis, explained* (August 18, 2020)

Link (pdf): https://www.dropbox.com/s/iychwnb6qge1aze/thesis13.pdf?dl=0

1PrimenumbersandtheRiemannzetafunction.byJørgenVeisdal

[17] ___

-1- John Derbyshire, *Prime Obsession: Berhard Riemann and the Greatest Unsolved Problem in Mathematics,* Plume (Mai 24, 2004). ISBN-13: 978-0452285255

-2- Oresme, Nicole (c. 1360). *Quaestiones super Geometriam Euclidis* [Questions concerning Euclid's Geometry], Franz Steiner Verlag; Bilingual Edition (December 30, 2010).

[18] ___ Ayoub, Raymond (1974). *Euler and the zeta function*. Amer. Math. Monthly. 81: 1067–86. doi:10.2307/2319041.

Link: https://www.maa.org/sites/default/files/pdf/upload_library/22/Ford/RaymondAyoub.pdf

[19] ___ Leonhard Euler. *Variae observationes circa series infinitas*. Commentarii academiae scientiarum Petropolitanae 9, 1744, pp. 160–188, Theorems 7 and 8.

Link: http://eulerarchive.maa.org/docs/originals/E072.pdf

[20] ___ David Wells, *Prime Numbers: The Most Mysterious Figures in Math*, p 72, Wiley (June 2005).

[21] ___ Vladimir Ryazanov, *A disproof of the Riemann hypothesis on zeros of ζ–function*, April 12, 2019

Link: https://arxiv.org/pdf/1808.10774.pdf

[22] ___ Bahattin Gunes. *The analysis of the Riemann hypothesis*. 2019. hal-02077752v1f.

Link: https://hal.archives-ouvertes.fr/hal-02077752v1/document

[23] ___ Pegg, Ed (2004), *Ten Trillion Zeta Zeros*, Math Games website. A discussion of Xavier Gourdon's calculation of the first ten trillion non-trivial zeros.

Link: http://mathpuzzle.com/MAA/28-Ten%20Trillion%20Zeta%20Zeroes/mathgames_10_18_04.html

228 References

[24] ___ Xavier Gourdon. *The 10¹³ first zeros of the Riemann zeta function, and zeros computation at very large height*, October 2004.

Links:

-1- http://numbers.computation.free.fr/Constants/Miscellaneous/zetazeros1e13-1e24.pdf

-2- http://www.mathpuzzle.com/MAA/28-Ten%20Trillion%20Zeta%20Zeroes/mathgames_10_18_04.html

[25] ___ Archived version of the ZetaGrid website: https://web.archive.org/web/20131005173705/http://www.zetagrid.net/

[26] ___ Weisstein, Eric W. "Skewes Number." From MathWorld--A Wolfram Web Resource. https://mathworld.wolfram.com/SkewesNumber.html

[27] ___ Hilbert, David (1902). *Mathematical Problems*. Bulletin of the American Mathematical Society. 8 (10): 437–479. doi:10.1090/S0002-9904-1902-00923-3. Earlier publications (in the original German) appeared in Hilbert, David (1900). "Mathematische Probleme". Göttinger Nachrichten: 253–297. and Hilbert, David (1901). Archiv der Mathematik und Physik. 3. 1: 44–63, 213–237.

Link: https://doi.org/10.1090%2FS0002-9904-1902-00923-3

[28] ___

-1- Borwein, P., Choi, S., Rooney, B., and Weirathmueller, A. (2008). *The Riemann Hypothesis*, p6, Canadian Mathematical Society.

-2- Ryan Dingman, *The Riemann Hypothesis,* March 12, 2010.

Link: https://wstein.org/edu/2010/414/projects/dingman.pdf

[29] ___ Julie Rehmeyer, *Michael Atiyah, Mathematician in Newton's Footsteps, Dies at 89,* The New York Times.

Link: https://www.nytimes.com/2019/01/11/obituaries/michael-atiyah-dead.html

[30] ___ Fields medal citation: Cartan, Henri (1968), "*L'oeuvre de Michael F. Atiyah*", Proceedings of International Conference of Mathematicians (Moscow, 1966), Izdatyel'stvo Mir, Moscow, pp. 9–14

[31] ___ The Abel Prize 2004.

Link: https://www.abelprize.no/c53865/seksjon/vis.html?tid=53873

[32] ___ HEIDELBERG LAUREATE FORUM FOUNDATION.

Link: https://www.newsroom.hlf-foundation.org/newsroom/lectures/video/lecture-the-riemann-hypothesis.html (retrieved August 12th, 2020).

[33] ___ Frankie Schembri, *Skeptism surrounds renowned mathematician's attempted proof of 160-year-old hypothesis,* Science (Magazine) Sep. 24, 2018

Link: https://www.sciencemag.org/news/2018/09/skepticism-surrounds-renowned-mathematician-s-attempted-proof-160-year-old-hypothesis

[34] ___ Gilead Amit, *Riemann Hypothesis likely remains unsolved despite claimed proof,* NewScienist, September 24, 2018.

Link: https://www.newscientist.com/article/2180504-riemann-hypothesis-likely-remains-unsolved-despite-claimed-proof/

[35] ___ The Clay Mathematics Institute

Link: https://www.claymath.org/about-cmi/clay-mathematics-institute-overview-and-history

Further readings:

Useful resources to read more about the Riemann Hypothesis:

https://encyclopediaofmath.org/index.php?title=Zeta-function

https://mathshistory.st-andrews.ac.uk/Biographies/Riemann

https://mathworld.wolfram.com/RiemannHypothesis.html

W.K.Clifford, *On the hypotheses which lie at the foundation of geometry* (1868), translated, Nature 8 1873 183 – reprinted in Clifford's Collected Mathematical Papers, London 1882 (MacMillan); New York 1968 (Chelsea) http://www.emis.de/classics/Riemann. Also in Ewald, William B., ed., 1996 *"From Kant to Hilbert: A Source Book in the Foundations of Mathematics"*, 2 vols. Oxford Uni. Press: 652–61.

Chapter 12 : Primes in Arithmetic Progressions

[1] ___

-1- van der Corput, J. G. *Über Summen von Primzahlen und Primzahlquadraten.* Math. Ann. 116, 1-50, 1939.

References

-2- S. Chowla, "There exists an infinity of 3--combinations of primes in A. P.," Proc. Lahore Phil. Soc., 6 (1944) 15--16. MR 7,2431

[2] ___ Weisstein, Eric W. "Prime Arithmetic Progression." From MathWorld--A Wolfram Web Resource.

Link: https://mathworld.wolfram.com/PrimeArithmeticProgression.html

[3] ___ Leonard Eugene Dickson, *History of the Theory of Numbers, Volume I: Divisibility and Primality*, Dover Publications; Illustrated Edition (June 3, 2005).

[4] ___ Green, Ben; Tao, Terence (2008), *The primes contain arbitrarily long arithmetic progressions*, Annals of Mathematics, 167 (2): 481–547, arXiv:math.NT/0404188, doi:10.4007/annals.2008.167.481, MR 2415379

Link: https://arxiv.org/pdf/math/0404188.pdf

[5] ___ From H. J. Weber, *Less Regular Exceptional and Repeating Prime Number Multiplets*, arXiv:1105.4092, Sect.3.

Link: https://arxiv.org/abs/1105.4092

[6] ___ Jens Kruse Andersen, Primes in Arithmetic Progression Records.

Link: http://primerecords.dk/aprecords.htm#2019

[7] ___ H. Dubner, T. Forbes, N. Lygeros, M. Mizony, H. Nelson, P. Zimmermann, *Ten consecutive primes in arithmetic progression*, Mathematics of Computation 71 (2002), 1323–1328.

Link: https://www.ams.org/journals/mcom/2002-71-239/S0025-5718-01-01374-6/S0025-5718-01-01374-6.pdf

[8] ___ GRANVILLE, Andrew. *Prime number patterns*. The American Mathematical Monthly, 2008, vol. 115, no 4, p. 279-296.

Link: https://www.maa.org/sites/default/files/pdf/upload_library/22/Ford/Granville.pdf

[9] ___ Online Encyclopedia for Integer Sequences

Link: https://oeis.org/A033188

Further readings

H. J. Weber, *Exceptional Prime Number Twins, Triplets and Multiplets*, November 11, 2018.

Link: https://arxiv.org/pdf/1102.3075.pdf

H. J. Weber, *Less Regular Exceptional and Repeating Prime Number Multiplets,* November 21, 2018

Link: https://arxiv.org/pdf/1105.4092.pdf

H. J. Weber, *Regularities of Twin, Triplet and Multiplet Prime Numbers*, November 6, 2018

Link: https://arxiv.org/pdf/1103.0447.pdf

Jarosław Wróblewski, *How to search for 26 primes in arithmetic progression?,* AP26 search, Version 3 (May 23, 2008).

Link: http://www.math.uni.wroc.pl/~jwr/AP26/AP26v3.pdf

Chapter 13 : Dirichlet's Theorem

[1] ___ Caldwell, C. K. "Prime Pages. Dirichlet's Theorem."
https://primes.utm.edu/glossary/page.php?sort=DirichletsTheorem

[2] ___ Carl Friedrich Gauss, *Disquisitiones Arithmeticae* (Leipzig, (Germany): Gerhard Fleischer, Jr., 1801), Section 297, pp. 507–508. From pp. 507–508:

Link: https://babel.hathitrust.org/cgi/pt?id=nyp.33433070725894&view=1up&seq=528

[3] ___ Derbyshire, J. *Prime Obsession: Bernhard Riemann and the Greatest Unsolved Problem in Mathematics.* New York: Penguin, pp. 95-97, 2004.

[4] ___ Weisstein, Eric W. "Dirichlet L-Series." From MathWorld--A Wolfram Web Resource.

Link: https://mathworld.wolfram.com/DirichletL-Series.html

[5] ___ Weisstein, Eric W. "Dirichlet's Theorem." From MathWorld--A Wolfram Web Resource.

Link: https://mathworld.wolfram.com/DirichletsTheorem.html

[6] ___

-1- Shanks, D. *Solved and Unsolved Problems in Number Theory*, 4th ed. New York: Chelsea, pp. 22-23, 1993.

232 References

-2- Weisstein, Eric W. "Dirichlet L-Series." From MathWorld--A Wolfram Web Resource.

Link: https://mathworld.wolfram.com/DirichletL-Series.html

[7] ___

-1- Linnik, Yu. V. (1944). *On the least prime in an arithmetic progression I. The basic theorem.* Rec. Math. (Mat. Sbornik) N.S. 15 (57): 139–178. MR 0012111.

Link: https://mathscinet.ams.org/mathscinet-getitem?mr=0012111

-2- Linnik, Yu. V. (1944). *On the least prime in an arithmetic progression II. The Deuring-Heilbronn phenomenon.* Rec. Math. (Mat. Sbornik) N.S. 15 (57): 347–368. MR 0012112.

Link: https://mathscinet.ams.org/mathscinet-getitem?mr=0012112

[8] ___ Xylouris, Triantafyllos (2011). *Über die Nullstellen der Dirichletschen L-Funktionen und die kleinste Primzahl in einer arithmetischen Progression [The zeros of Dirichlet L-functions and the least prime in an arithmetic progression]* (Dissertation for the degree of Doctor of Mathematics and Natural Sciences) (in German). Bonn: Universität Bonn, Mathematisches Institut. MR 3086819.

Link: https://mathscinet.ams.org/mathscinet-getitem?mr=3086819

[9] ___ Ribenboim, P. *The New Book of Prime Number Records.* New York: Springer-Verlag, 1989.

[10] ___ Leonard Eugene Dickson, *History Of The Theory Of Numbers: Divisibility And Primality*, Dover Publications Inc. (June 2005) ISBN-13: 978-0486442327.

[11] ___ David Wells, *Prime Numbers: The Most Mysterious Figures in Math,* p 42, Wiley (June 2005).

[12] ___ Ribenboim, Paulo, *The new book of prime number records. Springer-Verlag, New York,* 1996. xxiv+541 pp. ISBN: 0-387-94457-5.

[13] ___ A. Schinzel and W. Sierpinski, *Sur certaines hypotheses concernment les nombres premiers,* Acta. Arith., 4 (1958) 185-208. Erratum 5 (1958).

[14] ___ Caldwell, C. K. "Prime Pages. Hypothesis H."

Link: https://primes.utm.edu/glossary/page.php?sort=HypothesisH

References

[15] ___ Online Encyclopedia for Integer Sequences.

Link: https://oeis.org/A005574

[16] ___ Online Encyclopedia for Integer Sequences.

Link: https://oeis.org/A002496

[17] ___ Iwaniec, H. (1978). *Almost-primes represented by quadratic polynomials*. Inventiones Mathematicae. 47 (2): 178–188. Bibcode:1978InMat..47..171I. doi:10.1007/BF01578070.

Link: https://ui.adsabs.harvard.edu/abs/1978InMat..47..171I/abstract

[18] ___ Robert J. Lemke Oliver (2012). "Almost-primes represented by quadratic polynomials" (PDF). Acta Arithmetica. 151 (3): 241–261. doi:10.4064/aa151-3-2.

Link: https://doi.org/10.4064%2Faa151-3-2

[19] ___ Crandall, Richard; Pomerance, Carl B. (2005). *Prime Numbers: A Computational Perspective* (Second ed.). New York: Springer-Verlag. ISBN 0-387-25282-7. MR 2156291.

[20] ___ P. T. Bateman and R. A. Horn, "A heuristic asymptotic formula concerning the distribution of prime numbers," Math. Comp., 16 (1962) 363-367. MR 26:6139

Link: http://www.ams.org/mathscinet-getitem?mr=26:6139

Further readings

G. H. Hardy, Edward M. Wright, Andrew Wiles, Roger Heath-Brown, Joseph Silverman, *An Introduction to the Theory of Numbers,* 6th Edition, Oxford University Press (September 15, 2008).

Zijian Wang, *Elementary proof of Dirichlet theorem,* August 28, 2017.

Link: http://math.uchicago.edu/~may/REU2017/REUPapers/WangZijian.pdf

Chapter 14 : Formulas for Primes

[1] ___ P. Ribenboim, *The new book of prime number records*, 3rd edition, Springer-Verlag, New York, NY, 1995. pp. xxiv+541, ISBN 0-387-94457-5.

References

[2] ___ Caldwell, C. K. "Prime Pages. Is there a formula for the nth Prime?."

Link: https://primes.utm.edu/notes/faq/p_n.html

[3] ___ W. H. Mills, *A prime-representing function*. Bulletin of the American Mathematical Society, 58,1951, 616-618.

Link: https://projecteuclid.org/download/pdf_1/euclid.bams/1183510803

[4] ___ Caldwell, C. K. and Cheng, Y. *Determining Mills' Constant and a Note on Honaker's Problem*. J. Integer Sequences 8, Article 05.4.1, 1-9, 2005.

Link: https://www.cs.uwaterloo.ca/journals/JIS/VOL8/Caldwell/caldwell78.html.

[5] ___ Online Encyclopedia for Integer Sequences.

Link: https://oeis.org/A051021

[6] ___ Online Encyclopedia for Integer Sequences.

Link: https://oeis.org/A051254/list

[7] ___ Tóth, László (2017), *A Variation on Mills-Like Prime-Representing Functions* (PDF), Journal of Integer Sequences, 20: 17.9.8, arXiv:1801.08014.

Link: https://cs.uwaterloo.ca/journals/JIS/VOL20/Toth2/toth32.pdf

[8] ___ David Wells, " Euler Quadratic", *Prime Numbers: The Most Mysterious Figures in Math*, p 77, Wiley (June 2005).

[9] ___ Online Encyclopedia for Integer Sequences.

Link: https://oeis.org/A007634

[10] ___

-1- J. H. Conway and R. K. Guy, *The Book of Numbers*, Copernicus Press, NY, 1996, p. 224-226.

-2- J.-M. De Koninck, Ces nombres qui nous fascinent, Entry 41, p. 16, Ellipses, Paris 2008.

-3- Le Lionnais, F. *Les nombres remarquables*. Paris: Hermann, pp. 88 and 144, 1983.

[11] ___ Weisstein, Eric W. "Prime-Generating Polynomial." From MathWorld--A Wolfram Web Resource.

Link: https://mathworld.wolfram.com/Prime-GeneratingPolynomial.html

Chapter 15 : The Goldbach Conjecture

[1] ___ Guy, Richard K. (2004). *Unsolved problems in number theory* (3rd ed.),p159, Springer-Verlag. ISBN 978-0-387-20860-2.

[2] ___

-1- Adolf Juskevic, Judith Kopelevic: *Christian Goldbach* 1690-1764 (Vita Mathematica), Birkhäuser Publishing House, 1994, ISBN 3764326786, pg. XII.

-2- The Editors of Encyclopaedia Britannica, *Christian Goldbach*, Encyclopædia Britannica, Encyclopædia Britannica, inc. (March 14, 2020)

Link: https://www.britannica.com/biography/Christian-Goldbach

[3] ___ David Wells, *Prime Numbers: The Most Mysterious Figures in Math*, pp 117-119, Wiley (June 2005).

[4] ___ Paul Heinrich Fuss, *Correspondance mathématique et physique de quelques célèbres géomètres du XVIIIème siècle: précédé d'une notice sur les travaux de Léonard Euler, tant imprimés qu'inédits et publiée sous les auspices de l'Académie impériale des sciences de Saint-Pétersbourg, Volume 2,* pp 125-129, l'Académie impériale des sciences, 1843.

Link (The Euler Archive):
http://eulerarchive.maa.org//correspondence/letters/OO0765

[5] ___ L. E. Dickson, *History of the theory of numbers* (Vol. 1, p421), Carnegie Institute of Washington, 1919. Reprinted by Chelsea Publishing, New York, 1971.

[6] ___ Paul Heinrich Fuss, *Correspondance mathématique et physique de quelques célèbres géomètres du XVIIIème siècle: précédé d'une notice sur les travaux de Léonard Euler, tant imprimés qu'inédits et publiée sous les auspices de l'Académie impériale des sciences de Saint-Pétersbourg,* pp 130-136, l'Académie impériale des sciences, 1843.

[7] ___ Weisstein, Eric W. "Goldbach Conjecture." From MathWorld--A Wolfram Web Resource.
Link: https://mathworld.wolfram.com/GoldbachConjecture.html

[8] ___ Hilbert, David (1902). "*Mathematical Problems*". Bulletin of the American Mathematical Society. 8 (10): 437–479. doi:10.1090/S0002-9904-1902-00923-3. Earlier publications (in the original German) appeared in Hilbert, David (1900). "*Mathematische Probleme*". Göttinger Nachrichten: 253–297. and Hilbert, David (1901). Archiv der Mathematik und Physik. 3. 1: 44–63, 213–237.

Link: https://www.ams.org/journals/bull/1902-08-10/S0002-9904-1902-00923-3/S0002-9904-1902-00923-3.pdf

236 References

[9] ___ E. LANDAU, *Gelöste und ungelöste Probleme aus der Theorie dr Primzahlverteilung und der Riemannschen Zetafunktion*, Proceedings of the fifth International Congress of Mathematicians, Cambridge, 1912, vol. 1, pp. 93—108 (p. 105). This address was reprinted in the Jahresbericht der Deutschen Math:Vereinigung, vol. 21 (1912), pp. 208—228.

[10] ___ G. H. Hardy and J. E. Littlewood, "*Some problems of `partitio numerorum': III: on the expression of a number as a sum of primes,*" Acta Math., 44 (1923) 1-70. Reprinted in "*Collected Papers of G. H. Hardy,*" Vol. I, pp. 561-630, Clarendon Press, Oxford, 1966.

[11] ___ I. M. Vinogradov, "*Representation of an odd number as the sum of three primes,*" Dokl. Akad. Nauk SSSR, 16 (1937) 179--195. Russian. [Proves that the odd Goldbach conjecture holds for sufficiently all large integers n]

[12] ___

-1- Chen, Jing Run and Wang, Tian Ze, "*The Goldbach problem for odd numbers,*" Acta Math. Sinica (Chin. Ser.), 39:2 (1996) 169--174. MR1411958.

Link: https://mathscinet.ams.org/mathscinet-getitem?mr=MR1411958

-2- Caldwell, C. K. "Prime Pages. Odd Goldbach Conjecture."

Link: https://primes.utm.edu/glossary/page.php?sort=OddGoldbachConjecture

[13] ___ J.-M. Deshouillers, G. Effinger,H. Te Riele, and D. Zinoviev, *A complete Vinogradov 3-primes theorem under the Riemann hypothesis*. In: Electronic Research Announcements of the AMS. Volume 3, 1997, pp. 99–104, 17. September 1997.

Link: https://www.ams.org/journals/era/1997-03-15/S1079-6762-97-00031-0/S1079-6762-97-00031-0.pdf

[14] ___ Tao, Terence (2014). "*Every odd number greater than 1 is the sum of at most five primes*". Math. Comp. 83 (286): 997–1038. arXiv:1201.6656. doi:10.1090/S0025-5718-2013-02733-0. MR 3143702.

Link: https://arxiv.org/pdf/1201.6656.pdf

[15] ___

-1- Helfgott, Harald A. (2015). "*The ternary Goldbach problem*". arXiv:1501.05438.

Link: https://arxiv.org/abs/1501.05438

-2- Helfgott, Harald A. (2013). "*Major arcs for Goldbach's theorem*". arXiv:1305.2897.

Link: https://arxiv.org/abs/1305.2897

-3- Helfgott, Harald A. (2012). "*Minor arcs for Goldbach's problem*". arXiv:1205.5252.

Link: https://arxiv.org/abs/1205.5252

[16] ___ Yamada, Tomohiro (2015-11-11). "Explicit Chen's theorem". arXiv:1511.03409.

Link: https://arxiv.org/abs/1511.03409

[17] ___ Montgomery, H. L.; Vaughan, R. C. (1975). "The exceptional set in Goldbach's problem" (PDF). Acta Arithmetica. 27: 353–370. doi:10.4064/aa-27-1-353-370.

Link: http://matwbn.icm.edu.pl/ksiazki/aa/aa27/aa27126.pdf

[18] ___ Tomás Oliveira e Silva, *Goldbach conjecture verification*

Link: http://sweet.ua.pt/tos/goldbach.html

[19] ___

-1- Weisstein, Eric W. "Goldbach Conjecture." From MathWorld--A Wolfram Web Resource.

Link: https://mathworld.wolfram.com/GoldbachConjecture.html

-2- Bilal Khan, Kiran R. Bhutani, *Sparse Periodic Goldbach Sets*.

Link: https://citeseerx.ist.psu.edu/viewdoc/download?doi=10.1.1.580.9679&rep=rep1&type=pdf

[20] ___

-1- A. Desboves. Nouv. Ann. Math., 14:293, 1855.

-2- Pipping, N. "*Die Goldbachsche Vermutung und der Goldbach-Vinogradovsche Satz*." Acta. Acad. Aboensis, Math. Phys. **11**, 4-25, 1938.

-3- M. L. Stein and P. R. Stein. Experimental results on additive 2 bases. BIT, 38:427–434, 1965b.

-4- I. M. Vinogradov. *Representation of an odd number as a sum of three primes*. Comptes Rendus (Doklady) de l'Acadmie des Sciences de l'U.R.S.S., 15:169–172, 1937.

References

-5- A. Granville, J. van der Lune, and H. te Riele. *Checking the Goldbach conjecture on a vector computer*. Number Theory and Applications: Proceedings of the NATO Advanced Study Institute, pages 423–433, April 27-May 5, 1988.

-6- M. K. Sinsalo. *Checking the Goldbach conjecture up to $4 \cdot 10^{11}$*. Math. Comput., 61:931–934, 1993.

-7- J.-M. Deshouillers, H. te Riele, and Y. Saouter. New experimental results concerning the Goldbach conjecture. Algorithmic Number Theory: Proceedings of the 3rd International Symposium, pages 204–215, June 21-25, 1998.

-8- J. Richstein. *Verifying the Goldbach conjecture up to $4 \cdot 10^{14}$*. Math. Comput., 70:1745–1750, 2001.

-9- T. O. e Silva. *Verification of the Goldbach conjecture up to $2 * 10^{16}$*, available at: http://listserv.nodak.edu/scripts/wa.exe?A2=ind0303&L=nmbrthry&P=2394

-10- T. O. e Silva. *Verification of the Goldbach conjecture up to $6 * 10^{16}$*, available at: http://listserv.nodak.edu/scripts/scripts/wa.exe?A2=ind0310&L=nmbrthry&P=168

-11- T. O. e Silva. *New Goldbach conjecture verification limit*, available at: http://listserv.nodak.edu/scripts/scripts/wa.exe?A2=ind0512&L=nmbrthry&T=0&P=3233

-12, 13, 14- Tomás Oliveira e Silva, *Goldbach conjecture verification*, available at: http://sweet.ua.pt/tos/goldbach.html

[21] ___ Li, H. "The Exceptional Set of Goldbach Numbers." Quart. J. Math. Oxford 50, 471-482, 1999.

[22] ___ Online Encyclopedia for Integer Sequences

Link: https://oeis.org/A002375

[23] ___ Subhash Kak, *Goldbach Partitions and Sequences*, Resonance Magazine, Volume 19, Issue 11, November 2014, pp 1028-1037.

Link: https://www.ias.ac.in/article/fulltext/reso/019/11/1028-1037

[24] ___

-1- Yuan Wang,*The Goldbach conjecture*, 2nd Edition, Series in Pure Mathematics: Volume 4, World Scientific Publishing Company (2003). ISBN: 978-981-238-159-0

-2- Halberstam, H. and Richert, H.-E. *Sieve methods (L.M.S. monographs)*. New York: Academic Press, (January 1, 1974) ISBN-13: 978-0123182500.

References

[25] ___ Dickson, Leonard E. (1971). *History of the Theory of Numbers* (4 volumes). 1. S.l.: Chelsea. p. 424. ISBN 0-8284-0086-5.

[26] ___

-1- Corbit, D. "Conjecture on Odd Numbers." sci.math posting. Nov 19, 1999.

Link: https://groups-beta.google.com/group/sci.math/msg/539c96e47e3ed582?hl=en&.

-2- Dudley, Daniel and Weisstein, Eric W. "Levy's Conjecture." From MathWorld--A Wolfram Web Resource.

Link: https://mathworld.wolfram.com/LevysConjecture.html

[27] ___ "Lemoine's Conjecture Verified to 10^10". June 19, 2019. Retrieved June 19, 2019.

Link: https://www.makethebrainhappy.com/2019/06/lemoines-conjecture-verified-to-1010.html

(Please note that this is not a reliable resource.)

[28] ___ Margenstern, M. (1984). "Results and conjectures about practical numbers". Comptes rendus de l'Académie des Sciences. 299: 895–898.

[29] ___ Melfi, G. (1996). "On two conjectures about practical numbers". Journal of Number Theory. 56: 205–210. doi:10.1006/jnth.1996.0012.

Link: https://doi.org/10.1006%2Fjnth.1996.0012

[30] ___ Margenstern, Maurice (1991), "Les nombres pratiques: théorie, observations et conjectures", Journal of Number Theory, 37 (1): 1–36, doi:10.1016/S0022-314X(05)80022-8, MR 1089787

Link: https://www.sciencedirect.com/science/article/pii/S0022314X05800228

[31] ___ Sigler, Laurence E. (trans.) (2002), Fibonacci's Liber Abaci, Springer-Verlag, pp. 119–121, ISBN 0-387-95419-8

[32] ___ Srinivasan, A. K. (1948), "Practical numbers" (PDF), Current Science, 17: 179–180, MR 0027799.

Link: https://www.currentscience.ac.in/Downloads/articleid01706017901800.pdf

[33] ___ Allyn Jackson, "Book Review: *Uncle Petros and Goldbach's Conjecture and The Wild Numbers*", Notices of the AMS, volume 47, number 10, November 2000.

240 References

Link: https://www.ams.org/notices/200010/rev-jackson.pdf

Further readings

Yuan Wang, *The Goldbach conjecture*, 2nd Edition, Series in Pure Mathematics: Volume 4, World Scientific Publishing Company (2003).pdf. ISBN: 978-981-238-159-0

Gove W. Effinger and David R. Hayes, *A complete solution to the polynomial 3-primes problem*, Bulletin (new series) of the American Mathematical Society, Volume 24, number 2, April 1991

Link: https://www.ams.org/journals/bull/1991-24-02/S0273-0979-1991-16035-0/S0273-0979-1991-16035-0.pdf

Jailton C. Ferreira, *On Goldbach's Conjecture*, September 2002.

Link: https://arxiv.org/pdf/math/0209232.pdf

Caldwell, C. K. "Prime Pages. Odd Goldbach Conjecture."

Link: https://primes.utm.edu/glossary/page.php?sort=OddGoldbachConjecture

David Quarel, *On a numerical upper bound for the extended Goldbach conjecture,* arXiv:1801.01813, December 29, 2017

Link: https://arxiv.org/abs/1801.01813

Hodges, Laurent. "*A Lesser-Known Goldbach Conjecture*." Mathematics Magazine 66, no. 1 (1993): 45-47. doi:10.2307/2690477.

Chapter 16 : Bertrand's postulate

[1] ___ Sondow, Jonathan and Weisstein, Eric W. "Bertrand's Postulate." From MathWorld--A Wolfram Web Resource.

Link: https://mathworld.wolfram.com/BertrandsPostulate.html

[2] ___ Bertrand, J. *Mémoire sur le nombre de valeurs que peut prendre une fonction quand on y permute les lettres qu'elle renferme.* J. l'École Roy. Polytech. 17, 123-140, 1845.

[3] ___ Caldwell, C. K. "Prime Pages. Bertrand's Postulate."

Link: https://primes.utm.edu/glossary/page.php?sort=BertrandsPostulate

References

[4] ___ Aigner, M. and Ziegler, G. M. *Proofs from the Book*, p 7, 2nd ed. New York: Springer-Verlag, 2000.

[5] ___ Online Encyclopedia for Integer Sequences

Link: https://oeis.org/A077463

[6] ___ Online Encyclopedia for Integer Sequences

Link: https://oeis.org/A060715

[7] ___ Chebyshev, P. *Mémoire sur les nombres premiers.* Mém. Acad. Sci. St. Pétersbourg 7, 17-33, (1850) 1854. Reprinted as §1-7 in Œuvres de P. L. Tschebychef, Tome I. St. Pétersbourg, Russia: Commissionaires de l'Academie Impériale des Sciences, pp. 51-64, 1899.

[8] ___ Ramanujan, S. *Collected Papers of Srinivasa Ramanujan* (Ed. G. H. Hardy, P. V. S. Aiyar, and B. M. Wilson). Providence, RI: Amer. Math. Soc., pp. 208-209, 2000.

[9] ___ Ramanujan, S. (1919). "A proof of Bertrand's postulate". Journal of the Indian Mathematical Society. 11: 181–182.

Link: https://www.imsc.res.in/~rao/ramanujan/CamUnivCpapers/Cpaper24/page2.htm

[10] ___ Online Encyclopedia for Integer Sequences

Link: https://oeis.org/A104272

[11] ___ Nagura, J (1952). *On the interval containing at least one prime number.* Proceedings of the Japan Academy, Series A. 28 (4): 177–181. doi:10.3792/pja/1195570997.

Link: https://projecteuclid.org/DPubS/Repository/1.0/Disseminate?handle=euclid.pja/1195570997&view=body&content-type=pdf_1

[12] ___ Lowell Schoenfeld (April 1976). *Sharper Bounds for the Chebyshev Functions $\theta(x)$ and $\psi(x)$, II.* Mathematics of Computation. 30 (134): 337–360. doi:10.2307/2005976. JSTOR 2005976.

Link: https://www.jstor.org/stable/2005976

[13] ___ R. L. Graham, D. E. Knuth and O. Patashnik, *Concrete Mathematics.* Addison-Wesley, Exercise 4.19.

[14] ___ M. El Bachraoui, *Primes in the Interval [2n, 3n]*, Int. J. Contemp. Math. Sci., Vol. 1, 2006, no. 13, 617 – 621.

242   References

Link: https://www.researchgate.net/publication/267076371

[15] ___ Hanson, Denis (1973), *On a theorem of Sylvester and Schur*, Canadian Mathematical Bulletin, 16 (2): 195–199, doi:10.4153/CMB-1973-035-3.

Link: https://doi.org/10.4153%2FCMB-1973-035-3

[16] ___ Hardy, G. H. and Wright, W. M. *Unsolved Problems Concerning Primes.* §2.8 and Appendix §3 in *An Introduction to the Theory of Numbers*, 5th ed. Oxford, England: Oxford University Press, pp. 19 and 415-416, 1979.

[17] ___ Ribenboim, P. *The New Book of Prime Number Records*, 3rd ed. New York: Springer-Verlag, pp. 397-398, 1996.

[18] ___ Weisstein, Eric W. "Landau's Problems." From MathWorld--A Wolfram Web Resource.

Link: https://mathworld.wolfram.com/LandausProblems.html

[19] ___ Weisstein, Eric W. "Brocard's Conjecture." From MathWorld--A Wolfram Web Resource.

Link: https://mathworld.wolfram.com/BrocardsConjecture.html

[20] ___ Online Encyclopedia for Integer Sequences.

Link: https://oeis.org/A050216

[21] ___

-1- Andrica, D. (1986). "Note on a conjecture in prime number theory". Studia Univ. Babes–Bolyai Math. 31 (4): 44–48. ISSN 0252-1938. Zbl 0623.10030.

-2- Dr. Dorin Andrica's website.

Link: http://www.dorinandrica.ro/

[22] ___ Guy, Richard K. (2004). *Unsolved problems in number theory* (3rd ed.). Springer-Verlag. ISBN 978-0-387-20860-2.

[23] ___

-1- David Wells, *Prime Numbers: The Most Mysterious Figures in Math*, John Wiley & Sons, Inc., 2005, p. 13.

-2- Imran Ghory's website.

Link: http://www.imranghory.org/

[24] ___

-1- Oppermann, L. (1882), "Om vor Kundskab om Primtallenes Mængde mellem givne Grændser", pp 169–179, Kongelige Danske Videnskabernes Selskabs (1883).

Link on Google Books:
https://books.google.tn/books?id=UQgXAAAAYAAJ&pg=PA169&rediresc=y

-2- Ribenboim, Paulo (2004), *The Little Book of Bigger Primes*, Springer, p. 183, ISBN 9780387201696.

[25] ___ David Wells, *Prime Numbers: The Most Mysterious Figures in Math*, John Wiley & Sons, Inc., 2005, p. 164.

Chapter 17 : Wilson's theorem

[1] ___ L. E. Dickson, *History of the theory of numbers* (Vol. 1, p59), Carnegie Institute of Washington, 1919. Reprinted by Chelsea Publishing, New York, 1971.

[2] ___ O'Connor, John J.; Robertson, Edmund F., *Abu Ali al-Hasan ibn al-Haytham*, MacTutor History of Mathematics archive, University of St Andrews.

Link: https://mathshistory.st-andrews.ac.uk/Biographies/Al-Haytham/

[3] ___ Joseph Louis Lagrange, *Demonstration d'un théorème nouveau concernant les nombres premiers* (Proof of a new theorem concerning prime numbers), Nouveaux Mémoires de l'Académie Royale des Sciences et Belles-Lettres (Berlin), vol. 2, pages 125–137 (1771).

Link: https://books.google.com/books?id=-UAAAAYAAJ&pg=PA125#v=onepage&q&f=false

[4] ___ Mackinnon, Nick (June 1987), *Prime Number Formulae*, The Mathematical Gazette, 71 (456): 113–114, doi:10.2307/3616496, JSTOR 3616496.

Link: http://www.jstor.org/stable/3616496

[5] ___ Caldwell, C. K. "Prime Pages. A proof of Wilson's Theorem."

Link: https://primes.utm.edu/notes/proofs/Wilsons.html

[6] ___ David Wells, *Prime Numbers: The Most Mysterious Figures in Math*, pp 237-238, Wiley (June 2005).

References

[7] ___ Agoh, Takashi; Dilcher, Karl; Skula, Ladislav, *Wilson quotients for composite moduli*. Math. Comp. 67 (1998), no. 222, 843–861.

Link: https://www.ams.org/journals/mcom/1998-67-222/S0025-5718-98-00951-X/S0025-5718-98-00951-X.pdf

[8] ___ Online Encyclopedia for Integer Sequences

Link: https://oeis.org/A007540

[9] ___ E. Costa, R Gerbicz, and D. Harvey, *A search for wilson primes*, arXiv:1209.3436.

Link: https://arxiv.org/pdf/1209.3436.pdf

[10] ___ Great Internet Mersenne Prime Search forum.

Link: https://www.mersenneforum.org/showthread.php?t=16028

[11] ___ Crandall, Richard; Dilcher, Karl; Pomerance, Carl, *A search for Wieferich and Wilson primes*. (p 446) *Math. Comp.* 66 (1997), no. 217, pp 433–449.

[12] ___

-1- Table adapted from Wilson prime, Wikipedia online encyclopedia.

-2- OEIS: https://oeis.org/A007540

-3- OEIS: https://oeis.org/A079853

Further readings

Fred G. Elston, A Generalization of Wilson's Theorem, Mathematics Magazine, Vol. 30, No. 3 (Jan. – Feb. 1957), pp. 159-162 (4 pages), Taylor & Francis, Ltd. DOI: 10.2307/3029318

Link: https://www.jstor.org/stable/3029318

Thomas Jeffery, *A Generalization of Wilson's Theorem,* A Thesis presented to The University of Guelph (Advisor: Dr. Rajesh Pereira), Guelph, Ontario, Canada (November 2018)

Link: https://atrium.lib.uoguelph.ca/xmlui/bitstream/handle/10214/14690/Jeffery_Thomas_201812_Msc.pdf?sequence=3&isAllowed=y

Vardi, I. *Computational Recreations in Mathematica*. Addison-Wesley Pub Co (April 1, 1991), p. 73.

Chapter 18 : The twin prime Conjecture

[1] ___ Guy, R. K. "Gaps between Primes. Twin Primes." §A8 in *Unsolved Problems in Number Theory*, 2nd ed. New York: Springer-Verlag, pp. 19-23, 1994.

[2] ___ Chen, J.-R. "*On the Representation of a Large Even Number as the Sum of a Prime and the Product of at Most Two Primes.*" Sci. Sinica **16**, 157-176, 1973.

[3] ___

-1- Online Encyclopedia for Integer Sequences.

Link: https://oeis.org/A077800

-2- Online Encyclopedia for Integer Sequences.

Link: https://oeis.org/ A007510

[4] ___ Caldwell, C. K. "Prime Database: $2996863034895 \cdot 2^{1\,290\,000} - 1$." Retrieved on Mai 15, 2020.

Link(1): https://primes.utm.edu/primes/page.php?id=122213

Link(2): https://primes.utm.edu/top20/page.php?id=1#records

[5] ___ Weisstein, Eric W. "Twin Primes." From MathWorld--A Wolfram Web Resource.

Link: https://mathworld.wolfram.com/TwinPrimes.html

[6] ___ Tietze, H. "Prime Numbers and Prime Twins." Ch. 1 in *Famous Problems of Mathematics: Solved and Unsolved Mathematics Problems from Antiquity to Modern Times*. New York: Graylock Press, pp. 1-20, 1965.

[7] ___

-1- de Polignac, A. (1849). "*Recherches nouvelles sur les nombres premiers*" [New research on prime numbers]. Comptes rendus (in French). 29: 397–401. From p. 400.

Link: https://babel.hathitrust.org/cgi/pt?id=mdp.39015035450967&view=1up&seq=411

-2- Dickson, L. E. History of the Theory of Numbers, Vol. 1: Divisibility and Primality. New York: Dover, 2005.

References

[8] ___

-1- Brun, V. (1915), "Über das Goldbachsche Gesetz und die Anzahl der Primzahlpaare", Archiv for Mathematik og Naturvidenskab (in German), 34 (8): 3–19, ISSN 0365-4524, JFM 45.0330.16.

-2- Ribenboim, P. "Twin Primes." §4.3 in The New Book of Prime Number Records. New York: Springer-Verlag, p. 201, 1996.

[9] ___ Sebah, Pascal; Gourdon, Xavier. "Introduction to twin primes and Brun's constant computation". CiteSeerX 10.1.1.464.1118.

Link: https://citeseerx.ist.psu.edu/viewdoc/summary?doi=10.1.1.464.1118

[10] ___ Euler, Leonhard (1737). "*Variae observationes circa series infinitas*" [Various observations concerning infinite series]. Commentarii Academiae Scientiarum Petropolitanae. 9: 160–188.

[11] ___ P. T. Bateman, Harold G. Diamond, *Analytic Number Theory: An Introductory Course*, World Scientific, (2004) pp.313, 334–335. ISBN 981-256-080-7.

[12] ___ Hardy, G. H. and Wright, E. M. *An Introduction to the Theory of Numbers*, 5th ed. Oxford, England: Clarendon Press, 1979.

[13] ___ Le Lionnais, F. *Les nombres remarquables*. Paris: Hermann, p. 49, 1983.

[14] ___ Zhang, Yitang (2014). "Bounded gaps between primes". Annals of Mathematics. 179 (3): 1121–1174. doi:10.4007/annals.2014.179.3.7. MR 3171761.

Link: https://doi.org/10.4007%2Fannals.2014.179.3.7

[15] ___ Maggie McKee, First proof that infinitely many prime numbers come in pairs, Nature Magazine, 14 May 2013. doi:10.1038/nature.2013.12989

Link: https://www.nature.com/news/first-proof-that-infinitely-many-prime-numbers-come-in-pairs-1.12989

[16] ___ Maynard, James (20 November 2013). "Small Gaps Between Primes". arXiv: 1311.4600.

Link: https://arxiv.org/abs/1311.4600

[17] ___ Bounded gaps between primes (Main page), Polymath8.

Link: https://asone.ai/polymath/index.php?title=Bounded_gaps_between_primes

[18] ___ Online Encyclopedia for Integer Sequences

References

Link (1): https://oeis.org/A023200

Link (2): https://oeis.org/A046132

[19] ___

-1- Weisstein, Eric W. "Cousin Primes." From MathWorld--A Wolfram Web Resource.

Link: https://mathworld.wolfram.com/CousinPrimes.html

-2- Segal, B. (1930). "Generalisation du théorème de Brun". C. R. Acad. Sci. URSS (in Russian). 1930: 501–507. JFM 57.1363.06.

Link: https://zbmath.org/?format=complete&q=an:57.1363.06

[20] ___ Weisstein, Eric W. "Sexy Primes." From MathWorld--A Wolfram Web Resource.

Link: https://mathworld.wolfram.com/SexyPrimes.html

[21] ___ Online Encyclopedia for Integer Sequences

Link (1): https://oeis.org/A023201

Link (2): https://oeis.org/A046117

[22] ___ Kaiser, Peter (May 2019). "sexy prime triplet". Mersenne forum. Retrieved February 27, 2020.

Link: https://www.mersenneforum.org/showpost.php?p=527198&postcount=37

Further readings

Wells, D. The Penguin Dictionary of Curious and Interesting Numbers. Middlesex, England: Penguin Books, p. 41, 1986.

A lecture by Terry Tao, Ph.D. Small and Large Gaps Between the Primes (2014).

Link: https://www.youtube.com/watch?v=pp06oGD4m00&t=425&ab_channel=UCLA

Burton, D. M. Elementary Number Theory, 4th ed. Boston, MA: Allyn and Bacon, 1989.

Ball, W. W. R. and Coxeter, H. S. M. Mathematical Recreations and Essays, 13th ed. New York: Dover, p. 64, 1987.

Chris K. Caldwell, *An amazing prime heuristic*, November 2000.

Link: https://www.utm.edu/staff/caldwell/preprints/Heuristics.pdf

Chapter 19 : Ulam spiral

[1] ___ Gardner, M. (March 1964), *Mathematical Games: The Remarkable Lore of the Prime Number*, Scientific American, 210: 120–128, doi:10.1038/scientificamerican0364-120

Link: https://doi.org/10.1038%2Fscientificamerican0364-120

[2] ___ Stein, M. L.; Ulam, S. M.; Wells, M. B. (1964), *A Visual Display of Some Properties of the Distribution of Primes*, American Mathematical Monthly, Mathematical Association of America, 71 (5): 516–520, doi:10.2307/2312588, JSTOR 2312588

Link: https://www.jstor.org/stable/2312588

[3] ___ Mollin, R.A. (1996), "Quadratic polynomials producing consecutive, distinct primes and class groups of complex quadratic fields" (PDF), Acta Arithmetica, 74: 17–300

Link: http://matwbn.icm.edu.pl/ksiazki/aa/aa74/aa7412.pdf

[4] ___ G.H. Hardy and J.E. Littlewood, *Partitio numerorum III: On the expression of a number as a sum of primes*, Acta Math. 44 (1923), 1–70.

Link: https://projecteuclid.org/download/pdf_1/euclid.acta/1485887559

[5] ___ Jacobson Jr., M. J.; Williams, H. C (2003), *New quadratic polynomials with high densities of prime values* (PDF), Mathematics of Computation, 72 (241): 499–519, doi:10.1090/S0025-5718-02-01418-7

Link: https://www.ams.org/journals/mcom/2003-72-241/S0025-5718-02-01418-7/S0025-5718-02-01418-7.pdf

[6] ___ Daus, P. H. (1932), *The March Meeting of the Southern California Section*, American Mathematical Monthly, Mathematical Association of America, 39 (7): 373-374, doi: 10.1080/00029890.1932.11987331, JSTOR: 2300380.

Link: https://www.jstor.org/stable/2300380

[7] ___ J. Barthel, P. Sgobba, and F. Zhu,(2015) *Visualising the distribution of primes*, Experimental Mathematics Lab Project, University of Luxembourg.

Link: http://math.uni.lu/eml/projects/reports/prime-distribution.pdf

Further readings

https://www.betweenartandscience.com/ulamspiral_words.html

http://empslocal.ex.ac.uk/people/staff/mrwatkin/zeta/ulam.htm

Dewdney, A. K. *Computer Recreations: How to Pan for Primes in Numerical Gravel.* Sci. Amer. 259, 120-123, July 1988.

Ellerstein, S. M., *The Pronic Renaissance: The Ulam Square Spiral.* J. Recr. Math. 29, 188-189, 1998.

Credits and references of figures and images

Note:

The figures and images that are not referanced are either created or recreated by the author.

Chapter 1:

Figure 1: Image in the public domain. From https://debuglies.com/
Figure 2: From www.magicicada.org
Figure 3: US Forest Service

Chapter 2:

Figure 1: Courtesy NASA/JPL-Caltech.
Figure 2: Arecibo Observatory, from Wikipedia, the free encyclopedia. (Image in the public domain)
Figure 3: Arecibo Message, from Wikipedia, the free encyclopedia. (Image in the public domain)
Figure 4: Retreived from *The Arecibo Message* on http://www.physics.utah.edu/~cassiday/p1080/lec08.html

Chapter 3:

Figure 1: Julius Caesar, *marble sculpture by Andrea di Pietro di Marco Ferrucci, c. 1512–14; in the Metropolitan Museum of Art, New York City.* The Metropolitan Museum of Art, New York; Bequest of Benjamin Altman, 1913, 14.40.676, www.metmuseum.org
Figure 2: Caesar cipher, from Wikipedia, the free encyclopedia. (Image in the public domain)
Figure 3: Chuck Painter / Stanford News Service

Credits and references of figures and images

Chapter 5:

Figure 1: Euclid, from Wikipedia, the free encyclopedia. (Image in the public domain)
Figure 2: Carl Friedrich Gauss, from Wikipedia, the free encyclopedia. (Image in the public domain)

Chapter 6:

Figure 1: Eratosthenes, from Wikipedia, the free encyclopedia. (Image in the public domain)
Figure 4: Leonhard Euler, from Wikipedia, the free encyclopedia (the French version). (Image in the public domain)

Chapter 8:

Figure 1: Euler's totient function, from Wikipedia, the free encyclopedia. (Image in the public domain)

Chapter 9:

Figure 1: Marin Mersenne, from Wikipedia, the free encyclopedia. (Image in the public domain)
Figure 2: Retreived from Aforismi.meglio.it
https://aforismi.meglio.it/aforisma.htm?id=6c84
Figure 3: D. H. Lehmer, from Wikipedia, the free encyclopedia. (Image in the public domain)
Figure 4: Édouard Lucas, from Wikipedia, the free encyclopedia. (Image in the public domain)
Figure 6: Niccolò Fontana Tartaglia, from Wikipedia, the free encyclopedia. (Image in the public domain)

Chapter 10:

Fermat's picture: Pierre de Fermat, from Wikipedia, the free encyclopedia. (Image in the public domain)
Figure 5: Octahedral numbe, from Wikipedia, the free encyclopedia. (Image in the public domain)
Figure 6: Andrew Wiles, from Wikipedia, the free encyclopedia. (Image in the public domain)

Chapter 11:

Riemann's picture: Bernhard Riemann, from Wikipedia, the free encyclopedia. (Image in the public domain)
Figure 1: On the Number of Primes Less Than a Given Magnitude, from Wikipedia, the free encyclopedia. (Image in the public domain)
Figure 2: The spiral of Theodorus, from Wikipedia, the free encyclopedia. (Image in the public domain)
Figure 4: From mathspadilla.com (Image in the public domain)
http://www.mathspadilla.com/matI/Unit1-RealNumbers/conjuntos_reales.png

Chapter 12:

Figure 1: Edward Waring, from Wikipedia, the free encyclopedia. (Image in the public domain)
Figure 2: Joseph-Louis Lagrange, from Wikipedia, the free encyclopedia. (Image in the public domain)
Figure3: Terence Tao, (adapted) from Wikipedia, the free encyclopedia. (Image in the public domain)
Figure 4: Ben Joseph Green, from Wikipedia, the free encyclopedia. (Image in the public domain)

Chapter 15:

Figure 1: Goldbach's conjecture, from Wikipedia, the free encyclopedia. (Image in the public domain)
Figure 2: Goldbach's conjecture, from Wikipedia, the free encyclopedia. (Image in the public domain)

Chapter 16:

Figure 1: Pafnuty Lvovic, from Wikipedia, the free encyclopedia. (Image in the public domain)
Figure 2: Joseph-Louis Bertrand, from Wikipedia, the free encyclopedia. (Image in the public domain)
Figure3: Srinivasa Ramanujan, from Wikipedia, the free encyclopedia. (Image in the public domain)
Figure 4: Paul Erdős, from Wikipedia, the free encyclopedia. (Image in the public domain)
Figure 5: Andrica's conjecture, from Wikipedia, the free encyclopedia. (Image in the public domain)

Credits and references of figures and images

Chapter 17:

Figure 1: John Wilson, from Wikipedia, the free encyclopedia. (Image in the public domain)

Figure 2: Hasan Ibn al-Haytham, from Wikipedia, the free encyclopedia. (Image in the public domain)

Chapter 19:

Figure 1: Stanisław Ulam, from Wikipedia, the free encyclopedia. (Image in the public domain)

Figure 2: The Institution for Science Advancement, http://ifsa.my/articles/the-prime-spiral-between-art-science

Figure3: Ulam Spiral, from Wikipedia, the free encyclopedia. (Image in the public domain)

Figure 4: Ulam Spiral, from Wikipedia, the free encyclopedia. (Image in the public domain)

Appendix I :
List of the first 1000 prime numbers

2	3	5	7	11	13	17	19
23	29	31	37	41	43	47	53
59	61	67	71	73	79	83	89
97	101	103	107	109	113	127	131
137	139	149	151	157	163	167	173
179	181	191	193	197	199	211	223
227	229	233	239	241	251	257	263
269	271	277	281	283	293	307	311
313	317	331	337	347	349	353	359
367	373	379	383	389	397	401	409
419	421	431	433	439	443	449	457
461	463	467	479	487	491	499	503
509	521	523	541	547	557	563	569
571	577	587	593	599	601	607	613
617	619	631	641	643	647	653	659
661	673	677	683	691	701	709	719
727	733	739	743	751	757	761	769
773	787	797	809	811	821	823	827

Credits and references of figures and images 255

829	839	853	857	859	863	877	881
883	887	907	911	919	929	937	941
947	953	967	971	977	983	991	997
1009	1013	1019	1021	1031	1033	1039	1049
1051	1061	1063	1069	1087	1091	1093	1097
1103	1109	1117	1123	1129	1151	1153	1163
1171	1181	1187	1193	1201	1213	1217	1223
1229	1231	1237	1249	1259	1277	1279	1283
1289	1291	1297	1301	1303	1307	1319	1321
1327	1361	1367	1373	1381	1399	1409	1423
1427	1429	1433	1439	1447	1451	1453	1459
1471	1481	1483	1487	1489	1493	1499	1511
1523	1531	1543	1549	1553	1559	1567	1571
1579	1583	1597	1601	1607	1609	1613	1619
1621	1627	1637	1657	1663	1667	1669	1693
1697	1699	1709	1721	1723	1733	1741	1747
1753	1759	1777	1783	1787	1789	1801	1811
1823	1831	1847	1861	1867	1871	1873	1877
1879	1889	1901	1907	1913	1931	1933	1949
1951	1973	1979	1987	1993	1997	1999	2003
2011	2017	2027	2029	2039	2053	2063	2069
2081	2083	2087	2089	2099	2111	2113	2129
2131	2137	2141	2143	2153	2161	2179	2203
2207	2213	2221	2237	2239	2243	2251	2267

2269 2273 2281 2287 2293 2297 2309 2311
2333 2339 2341 2347 2351 2357 2371 2377
2381 2383 2389 2393 2399 2411 2417 2423
2437 2441 2447 2459 2467 2473 2477 2503
2521 2531 2539 2543 2549 2551 2557 2579
2591 2593 2609 2617 2621 2633 2647 2657
2659 2663 2671 2677 2683 2687 2689 2693
2699 2707 2711 2713 2719 2729 2731 2741
2749 2753 2767 2777 2789 2791 2797 2801
2803 2819 2833 2837 2843 2851 2857 2861
2879 2887 2897 2903 2909 2917 2927 2939
2953 2957 2963 2969 2971 2999 3001 3011
3019 3023 3037 3041 3049 3061 3067 3079
3083 3089 3109 3119 3121 3137 3163 3167
3169 3181 3187 3191 3203 3209 3217 3221
3229 3251 3253 3257 3259 3271 3299 3301
3307 3313 3319 3323 3329 3331 3343 3347
3359 3361 3371 3373 3389 3391 3407 3413
3433 3449 3457 3461 3463 3467 3469 3491
3499 3511 3517 3527 3529 3533 3539 3541
3547 3557 3559 3571 3581 3583 3593 3607
3613 3617 3623 3631 3637 3643 3659 3671
3673 3677 3691 3697 3701 3709 3719 3727
3733 3739 3761 3767 3769 3779 3793 3797

Credits and references of figures and images 257

3803 3821 3823 3833 3847 3851 3853 3863
3877 3881 3889 3907 3911 3917 3919 3923
3929 3931 3943 3947 3967 3989 4001 4003
4007 4013 4019 4021 4027 4049 4051 4057
4073 4079 4091 4093 4099 4111 4127 4129
4133 4139 4153 4157 4159 4177 4201 4211
4217 4219 4229 4231 4241 4243 4253 4259
4261 4271 4273 4283 4289 4297 4327 4337
4339 4349 4357 4363 4373 4391 4397 4409
4421 4423 4441 4447 4451 4457 4463 4481
4483 4493 4507 4513 4517 4519 4523 4547
4549 4561 4567 4583 4591 4597 4603 4621
4637 4639 4643 4649 4651 4657 4663 4673
4679 4691 4703 4721 4723 4729 4733 4751
4759 4783 4787 4789 4793 4799 4801 4813
4817 4831 4861 4871 4877 4889 4903 4909
4919 4931 4933 4937 4943 4951 4957 4967
4969 4973 4987 4993 4999 5003 5009 5011
5021 5023 5039 5051 5059 5077 5081 5087
5099 5101 5107 5113 5119 5147 5153 5167
5171 5179 5189 5197 5209 5227 5231 5233
5237 5261 5273 5279 5281 5297 5303 5309
5323 5333 5347 5351 5381 5387 5393 5399
5407 5413 5417 5419 5431 5437 5441 5443

5449 5471 5477 5479 5483 5501 5503 5507
5519 5521 5527 5531 5557 5563 5569 5573
5581 5591 5623 5639 5641 5647 5651 5653
5657 5659 5669 5683 5689 5693 5701 5711
5717 5737 5741 5743 5749 5779 5783 5791
5801 5807 5813 5821 5827 5839 5843 5849
5851 5857 5861 5867 5869 5879 5881 5897
5903 5923 5927 5939 5953 5981 5987 6007
6011 6029 6037 6043 6047 6053 6067 6073
6079 6089 6091 6101 6113 6121 6131 6133
6143 6151 6163 6173 6197 6199 6203 6211
6217 6221 6229 6247 6257 6263 6269 6271
6277 6287 6299 6301 6311 6317 6323 6329
6337 6343 6353 6359 6361 6367 6373 6379
6389 6397 6421 6427 6449 6451 6469 6473
6481 6491 6521 6529 6547 6551 6553 6563
6569 6571 6577 6581 6599 6607 6619 6637
6653 6659 6661 6673 6679 6689 6691 6701
6703 6709 6719 6733 6737 6761 6763 6779
6781 6791 6793 6803 6823 6827 6829 6833
6841 6857 6863 6869 6871 6883 6899 6907
6911 6917 6947 6949 6959 6961 6967 6971
6977 6983 6991 6997 7001 7013 7019 7027
7039 7043 7057 7069 7079 7103 7109 7121

7127 7129 7151 7159 7177 7187 7193 7207
7211 7213 7219 7229 7237 7243 7247 7253
7283 7297 7307 7309 7321 7331 7333 7349
7351 7369 7393 7411 7417 7433 7451 7457
7459 7477 7481 7487 7489 7499 7507 7517
7523 7529 7537 7541 7547 7549 7559 7561
7573 7577 7583 7589 7591 7603 7607 7621
7639 7643 7649 7669 7673 7681 7687 7691
7699 7703 7717 7723 7727 7741 7753 7757
7759 7789 7793 7817 7823 7829 7841 7853
7867 7873 7877 7879 7883 7901 7907 7919

Source:

https://primes.utm.edu/lists/small/1000.txt

If you want the list of the first 10,000 primes, you can find it here:

https://www.di-mgt.com.au/primes10000.txt

If this was not enough, here is the list of the first **1000 billion** primes:

http://compoasso.free.fr/primelistweb/page/prime/listeonlineen.php

Index

2

2018 Heidelberg Laureate Forum, 144

7

7 Millenium Prize Problems, 144

A

Abel Prize, 108, 144, 229, 239
Adrien-Marie Legendre, 63
Adelmann
 Leonard Adelmann, 37
Advanced Research Project Agency, 32
American Mathematical Society, 111, 221, 222, 223, 227, 230, 238, 244, 246, 250
analytic geometry, 97, 98
Andrica
 Dorin Andrica, 184, 236, 253
Andrica's conjecture, 184, 185, 187

Archimedes, 103
Arecibo Message, 25, 215
Arecibo radio telescope, 25
arithmetic sequences, 146, 149, 240
Arthur Oliver Lonsdale Atkin, 59
Asimov
 Isaac Asimov, 175
Asymmetric Cipher, 33
Atiyah
 Michael Atiyah, 144, 238

B

Bachraoui
 Mohamed El Bachraoui, 182
Baconian Cipher, 31
Basel Problem, 140
Bateman–Horn conjecture, 158
Beal
 Andrew Beal, 111
Beaumont-de-Lomagne, 97
Berlin Academy, 128, 132

Bernstein, Daniel Julius, 59
Bertrand
 Joseph Louis François Bertrand, 178
Bertrand's postulate, 177, 178, 179, 181
Bloomsbury Publishing, 175
Boklan
 Kent Boklan, 123
Bombelli
 Rafael Bombelli, 138
Bordeaux, 97
Brahmagupta, 99
Breselenz, 130
Breuil
 Christophe Breuil, 108
Brocard
 Pierre René J. B. H. Brocard, 184
Brocard's conjecture, 184, 186, 187
Brun
 Viggo Brun, 198
Bunyakovsky Conjecture, 158

262 Index

C

Caesar, Julius, 30

Cambridge, 169, 213, 216, 219, 228, 235, 246
Cardano, 138
Carmichael
    Robert Carmichael, 75, 76, 77, 113, 221, 223
    Crmichael numbers, 77
Cartesian Geometry, 98
Castres, 97
Catalan
    Eugène Charles Catalan, 110
Catalan's conjecture, 110
Cauchy
    Augustin-Louis Cauchy, 132, 138, 103
central binomial coefficient, 181
Chebyshev
    Pafnuty Lvovich Chebyshev, 179
Chebyshev's Theorem, 179
Chen prime, 195
Chen's Theorem, 170
Chen's Theorems, 170
Chermoni
    Raanan Chermoni, 151
chromosomes, 16, 17
cicadas
    Periodical Cicadas, 17
Cicada, 14, 17
Clarke
    Arthur C. Clarke, 163
Clay Mathematical Institute, 144
complex plane, 137, 138, 142
computational mathematics, 15
Computational number theory, 15
computer science, 15, 34, 37, 145
Conjecture F, 206
Conrad
    Brian Conrad, 108
Conway
    John Conway, 123
Critical strip, 142, 143
Cunningham chain, 82, 224
CUNY Queens College, 123

D

de Bessy
    Bernard Frénicle de Bessy, 112
de la Vallée Poussin, Charles Jean, 63
de Moivre
    Abraham de Moivre, 138
de Polignac
    Alphonse de Polignac, 198
de Polignac's Conjecture, 200
de Polignac's conjecture, 198
decryption, 33, 39
d'Espagnet
    Etienne d'Espagnet, 97
Defence Research Establishment Valcartier in Canada, 23
Descartes, 97, 138
    René Descartes, 97
Deshouillers
    Jean-Marc Deshouillers, 169
Diamond
    Fred Diamond, 108
Dickson's conjecture, 14, 155, 156
Dictionary.com, 49, 217
Diophantus, 98, 99, 108, 228
Dirichlet
    Johann Peter Gustav Lejeune Dirichlet, 131
    Peter Gustav Lejeune Dirichlet, 154
Dirichlet's Theorem on Primes in Arithmetic Progressions, 154
Double Wieferich Prime Pair, 117, 233
Doxiadis
    Apostolos Doxiadis, 175
Drake, Frank, 23
Drunkard's Walk, 52
DUBNER
    H. DUBNER, 151
Dumas, Stephane, 23

E

Effinger
    Gove Effinger, 169
Einstein
    Albert Einstein, 132
Elite High School of Kairouan, 9

encryption, 30, 31, 32, 33, 37, 39
ENIGMA, 31
Erdős, Paul, 76, 180
Eratosthenes, 14, 49, 50, 52, 53, 60, 68, 198, 217, 218
Euclid, 13, 14, 41, 44, 45, 48, 90, 92, 94, 113, 133, 216, 228, 237
   Elements, 41
Euclid of Alexandria, 13
Euclidean geometry, 131
Euler
   Leonhard Euler, 53, 59, 60, 69, 70, 71, 72, 74, 75, 76, 82, 85, 90, 94, 106, 114, 118, 119, 120, 121, 131, 138, 140, 141, 157, 163, 164, 166, 168, 198, 206, 212, 218, 220, 221, 222, 223, 224, 229, 231, 237, 244, 245, 246, 256
Evpatoria transmitter, 23

F

Faber and Faber Limited, 175
Fermat
   Fermat's Little Theorem, 73, 111, 113, 114, 115, 117
Fermat, 14, 73, 77, 81, 96, 97, 98, 99, 100, 101, 103, 105, 106, 107, 108, 109, 110, 111, 112, 113, 114, 115, 116, 117, 118, 119, 120, 121, 122, 123, 125, 126, 161, 166, 226, 227, 228, 229, 230, 231, 232, 233, 234, 235
Fermat Numbers, 118, 122, 123, 161, 233
Fields Medal, 144
Filip Saidac, 42
Forbes,
   Tony Forbes, 151
Ford, Kevin, 76
France, 97, 168, 227
Frey
   Gerhard Frey, 108
Frind
   Markus Frind, 151
Freudenthal, Hans, 23, 215
Fundamental Theorem of Arithmetic, 45, 217

G

Gardner
   Martin Gardner, 205
Gauss, 45, 64, 103, 125, 131, 143, 154, 155, 216, 217, 241
Germany, 128, 130, 144, 221, 241
Ghory
   Imran Ghory, 185, 253
Gimps
   Great Internet Mersenne Prime Search, 84
GIMPS, 84, 86, 88, 122, 224, 225
Girard
   Albert Girard, 105
globular star cluster M13, 25
Goldbach
   Christian Goldbach, 166, 245
Goldbach conjecture, 14, 173, 175, 246, 247, 248, 249, 250
Goldbach's Theorem, 120
Goldston
   Daniel Alan Goldston, 200
   Daniel Goldston, 149
Great Internet Mersenne Prime Search, 254
Green
   Ben Joseph Green, 149
Gronsfeld Cipher, 31
Guinness Book Of World Records, 107

H

Hadamard, Jacques, 63
Hanover, 130
Hanson
   Denis Hanson, 182
Hardy, 114, 165, 173, 199, 231, 243, 246, 251, 252, 256, 259
   Godfrey Harold Hardy, 160, 169, 206
Harmonic series, 139, 141
Heath-Brown
   Roger Heath-Brown, 148, 243
Helfgott, 247
   Harald Andrés Helfgott, 169

264 Index

Martin Hellmann, 32
Hilbert
  David Hilbert, 144, 168
  Hilbert's problems, 144
  Hilbert's Problems, 169
Hirzebruch
  Friedrich Ernst Peter Hirzebruch, 144
Hudalricus Regius, 79
Hypothesis H
  Conjecture, 156

I

Ibn-Alhaytham
  Hassan Ibn-Alhaytham, 189
Imaginary axis, 137, 142
International Congress of Mathematicians, 168, 169, 246
isolated primes, 195
Italy, 130
Ivory
  James Ivory, 114
Iwaniec
  Henryk Iwaniec, 158

J

Jingrun
  Chen Jingrun, 170, 195
Jobling
  Paul Jobling, 151

K

Kaliningrad, 166
Key Distribution problem, 32
Klauber
  Laurence Monroe Klauber, 209
Königsberg, 166
K-Theory, 144
Kummer
  Ernest Kummer, 109

L

Lagrange
  Joseph-Louis Lagrange, 103, 254, 149, 189
  Lagrange's Theorem, 191
Landau
  Edmund G. H. Landau, 169
  Edmund Landau, 182
  Landau's four-problem list, 158
  Landau's problems, 182
least common multiple, 19
Lee
  Gentry Lee, 163
Legendre, 66, 81, 131, 164, 182, 184, 186, 220
Lehmer
  Derrick Henry Lehmer, 77, 87
Leibniz
  Gottfried Leibniz, 98, 166
  Gottfried Wilhelm Leibniz, 113
Lemoine
  Émile Michel Hyacinthe Lemoine, 174
  Lemoine's Conjecture, 174
Levy
  Hyman Levy, 174

Linnik
  Yuri Vladimirovich Linnik, 155
  Linnik's constant, 155
  Linnik's theorem, 155
Littlewood
  John Edensor Littlewood, 169, 206
Los Alamos Scientific Laboratory, 204
L-series, 154
Lucas, 87, 89, 121
  Edouard Lucas, 89, 85, 120
Lucky Numbers, 164
Lygeros
  N. Lygeros, 151, 240

M

Magicicada, 17
Magicicadas, 18
Margenstern
  Maurice Margenstern, 174
Marin Mersenne, 79, 105
Merkle
  Ralph Merkle, 32
Maynard
  James Maynard, 200
Melfi
  Giuseppe Melfi, 174
Mengoli
  Pietro Mengoli, 140
Mersenne number, 78, 89
Mersenne Twister, 88, 225
Mihăilescu
  Preda V. Mihăilescu, 110
Mills, 161

Index **265**

William H. Mills, 161
Mills' constant, 162
Mills' Theorem, 163
Mirimanoff
   Dmitry Semionovitch Mirimanoff, 116
MIT Laboratory Of Computer Science, 37
Mizony
   M. Mizony, 151, 240
Montgomery
   Hugh Lowell Montgomery, 170
Multiplicative order, 82, 83

N

Nagura
   Jitsuro Nagura, 181
Nayan Hajratwala, 86, 88
Nelson
   H. Nelson, 151, 240
New Mexico, 204
Newton
   Isaac Newton, 98
non-trivial zero of R.H, 142
number line, 137

O

OEIS, 61, 62, 219, 230, 232, 233, 234, 241, 243, 244, 248, 251, 252, 254, 255, 257
Oliveira e Silva
   Tomás Oliveira e Silva, 170, 247, 248
Oppermann

Ludvig Henrik Ferdinand Oppermann, 186
Oresme
   Nicholas Oresme, 139
Orléans, 97

P

Paris, 168, 227, 230, 233, 245, 257
Parlement of Toulouse, 97
Pascal
   Blaise Pascal, 98
Pell's equations, 99
Perfect number, 89
Peter II, 166
Pietro Cataldi, 79, 85
Pintz
   János Pintz, 200
Pohl, Frederik, 52
Polignac's conjecture, 174
Pollock
   Frederick Pollock, 103
Pollock's conjectures, 103
Polymath project 8, 200
principle of least time, 98
Prussia, 166
Pseudo-Random Number Generator, 88
Puerto Rico, 25
Pythagorean Theorem, 133, 134

Q

Quadratic Residue, 82, 83

R

Ramanujan
   Srinivasa Ramanujan, 179, 251
real axis, 137, 142
real number, 60, 136, 137, 142, 161, 178, 198
Rhind Mathematical Papyrus, 13
Ribenboim,
   Paulo Ribenboim, 222, 235, 242, 243, 244, 252, 253, 256, 161, 231
Ribet
   Ken Ribet, 108
Riemann, 14, 63, 127, 128, 129, 130, 131, 132, 138, 139, 141, 142, 143, 144, 145, 162, 169, 235, 236, 237, 238, 239, 241, 246
Riemann Hypothesis, 14, 128, 132, 138, 162, 169, 235, 236, 237, 238, 239
Riemann Zeta function, 14
Riemannian geometry, 131, 132
Rockmore
   Dan Rockmore, 132
Ron Rivest, 37
RSA, 37, 40, 73
Ruiz
   Sebastián Martín Ruiz, 197
Russia, 166, 251
Russian Ministry of Foreign Affairs, 166

266 Index

S

S.P. Sundaram, 59
Safe primes, 81
Shamir
   Adi Shamir, 37
Schinzel
   Andrzej B.
      M.Schinzel, 155
   Andrzej Schinzel, 156
Schoenfeld
   Lowell Schoenfeld, 181, 252
Scientific American, 205, 213, 216, 258
Selasca, 130
Serre
   Jean-Pierre Serre, 108
Shimura
   Gorō Shimura, 108
Siegel
   Carl Ludwig Siegel, 130
Sierpinski
   Wacław Franciszek Sierpiński, 156
Sierpiński
   Sierpiński's conjecture, 76
   Wacław F. Sierpiński, 155
   Wacław Sierpiński, 160
sieve of Atkin, 59
Sophie Germain, 80, 81, 82, 109, 156, 224, 229
Sophie Germain primes, 82
spiral of Theodorus, 134
Stäckel
   Paul G. S. Stäckel, 198

Stanford University, 32
Stein
   Myron Stein, 204
Stieltjes
   Thomas Joannes Stieltjes, 144
Szemerédi's Theorem, 149

T

Taniyama
   Yutaka Taniyama, 108
Taniyama–Shimura conjecture, 108
Tartaglia
   Niccolò Fontana Tartaglia, 93
Taylor
   Richard Taylor, 108
te Riele
   Herman te Riele, 169
Tao, Terence, 169
Toth, 163
totient number, 74, 75
Tower of Hanoi, 89
Turing, Alan, 31
twin prime conjecture, 195, 198, 199, 200
twin Prime conjecture, 14

U

UK, 169, 175, 213
Ukraine, 23
Ulam
   Stanisław Marcin Ulam, 203
Ulam Spiral, 204, 205, 207, 212
Underwood
   Paul Underwood, 151

University of Göttingen, 131
University of Orléans., 97
University of Toulouse, 97
US Department Of Defense, 32
USA, 42, 204

V

van der Corput
   Johannes Gaultherus van der Corput, 148
Vaughan
   Robert Charles Vaughan, 170
Viginère Cipher, 31, 33
Vinogradov
   Ivan Matveevich Vinogradov, 169
von Neumann
   John von Neumann, 144
Voyager spacecraft, 22

W

Wantzel
   Pierre Wantzel, 125
Wells
   Mark Wells, 204
Wiles, Andrew, 107, 243
Whitefield Diffie, 32
Wieferich number, 116, 118
Wieferich primes, 116, 117
   Arthur Josef Wieferich, 116
Wilson
   John Wilson, 189

Wolf
  Marek Wolf, 201
Woltman, George, 84
World War II, 31
Wright, 161, 165, 199,
  231, 243, 252, 256
  Edward Maitland
    Wright, 160
Wroblewski

Jaroslaw
  Wroblewski, 151

Y

Yamada, 170
Yıldırım
  Cem Yalçın
    Yıldırım, 149, 200

Yvan Dutil, 23, 215

Z

Zhang
  Yitang Zhang, 199
Zimmermann
  P. Zimmermann,
    151, 240

www.ingramcontent.com/pod-product-compliance
Lightning Source LLC
Chambersburg PA
CBHW052343220526
45465CB00003BA/936